T0192837

SpringerBriefs in Mathematical Physics

Volume 40

SpringerBriefs are characterized in general by their size (50–125 pages) and fast production time (2–3 months compared to 6 months for a monograph).

Briefs are available in print but are intended as a primarily electronic publication to be included in Springer's e-book package.

Typical works might include:

- An extended survey of a field
- A link between new research papers published in journal articles
- A presentation of core concepts that doctoral students must understand in order to make independent contributions
- Lecture notes making a specialist topic accessible for non-specialist readers.

SpringerBriefs in Mathematical Physics showcase, in a compact format, topics of current relevance in the field of mathematical physics. Published titles will encompass all areas of theoretical and mathematical physics. This series is intended for mathematicians, physicists, and other scientists, as well as doctoral students in related areas.

More information about this series at http://www.springer.com/series/11953

Kentaroh Yoshida

Yang–Baxter Deformation of 2D Non-Linear Sigma Models

Towards Applications to AdS/CFT

 Springer

Kentaroh Yoshida
Department of Physics and Astronomy
Kyoto University
Kyoto, Japan

ISSN 2197-1757   ISSN 2197-1765 (electronic)
SpringerBriefs in Mathematical Physics
ISBN 978-981-16-1702-7   ISBN 978-981-16-1703-4 (eBook)
https://doi.org/10.1007/978-981-16-1703-4

This Springer imprint is published by the registered company Springer Nature Singapore Pte Ltd.
The registered company address is: 152 Beach Road, #21-01/04 Gateway East, Singapore 189721, Singapore

This book is dedicated to my parents, wife, and children in order to appreciate their understanding and patience during my work.

Preface

In recent years, research on integrability of nonlinear sigma models has attracted attention again, inspired by the discovery of integrable structures that exist behind the AdS/CFT correspondence. There are various directions for research on the integrable structures. Among them, the study of integrable deformations has attracted the interest of many researchers. In particular, there is a method for systematically performing integrable deformations, the so-called Yang–Baxter deformation, which is the main subject of this book.

The purpose of this book is to start with a pedagogical introduction to the classical integrability of nonlinear sigma models in two dimensions, and to give an elementary explanation of the Yang–Baxter deformation technique, which has gained popularity in recent research. There is a long history as to the integrability of the sigma model and there are many excellent standard textbooks. However, the Yang–Baxter deformation itself is a new subject and at least so far, there is no textbook thereof. Based on this situation, Chap. 1 is devoted to some basics on the classical integrability and the detail of the Yang–Baxter deformation is introduced carefully in Chap. 2.

Much progress has been made for the method of the Yang–Baxter deformation itself and the research is still vigorously continued. Hence, in order to motivate the readers who have learned Chaps. 1 and 2, Chapter 3 has been added to introduce what significant progress has been made in the study of String Theory by studying the Yang–Baxter deformation. This chapter is devoted to a concise overview for the recent progress on the Yang–Baxter deformation and the discovery of the generalized supergravity. Hence, it is quite different from the first two chapters and does not explain the contents in a pedagogical way and some technical terms are utilized without explaining the definition. The readers who are interested in the detail should read the references cited there.

I believe that writing this book, at this timing, will be helpful as a bridge for researchers currently working in String Theory and Classically Integrable System, and especially for young students who are interested in studying the Yang–Baxter deformation. It would be a pleasure to see if the readers of this book have made further progress in this area.

Kyoto, Japan   Kentaroh Yoshida

Acknowledgements

First of all, the author would like to express his gratitude to Professor Atsuo Kuniba for giving him the opportunity for writing this book. He also thanks Masayuki Nakamura, who has continued to encourage him in writing this book and has led to the completion of writing safely. In addition, he would like to appreciate the referee's comments and suggestions. These were really helpful to improve and polish the content of this book. Needless to say, the author is responsible for all errors and typos in this book.

He is grateful to Io Kawaguchi, who was one of his students, and Takuya Matsumoto for the collaboration on Yang–Baxter sigma models. In particular, the second chapter of this book is mainly based on the collaboration with them.

The first chapter is devoted to a pedagogical review of the classical integrability in nonlinear sigma models in two dimensions. The author would like to thank Marc Magro and Benoit Vicedo for significant discussions on the sigma-model integrability on some occasions. These were always valuable and led him to a deeper understanding of this subject.

He would like to thank Osamu Fukushima, Jun-ichi Sakamoto, and Yuta Sekiguchi for carefully reading the draft of this book. Discussions with them were very interesting and useful to finish this book better.

Contents

Chapter 1
Integrable Non-linear Sigma Models in (1+1)-Dimensions

Abstract This chapter will introduce the notion of classical integrability of non-linear sigma models in two dimensions. In general, non-linear sigma models are not classically integrable in general dimensions and even in two dimensions. However, there are well-known integrable examples in two dimensions such as principal chiral model and symmetric coset sigma model. Here for these two examples, the classical integrability will be described by explicitly presenting a classical Lax pair and the associated infinite-dimensional algebra, Yangian.

1.1 What is Classical Integrability?

The main subject of this book is concerned with the classical integrability of non-linear sigma models, and hence before going into the detail of the non-linear sigma model, it would be instructive to start from a pedagogical introduction of the notion of classical integrability itself.

In the undergraduate physics courses, students take a class on analytical mechanics and this would be the first time to learn the classical integrability. So, in the following, let us recall some elementary facts in the classical mechanical system with time t.

The Arnold-Liouville Theorem

Consider a classical (autonomous)[1] Hamiltonian system with the generalized coordinate $q = (q^1, \ldots, q^i, \ldots, q^N)$ and its canonical conjugate momentum $p = (p_1, \ldots, p_i, \ldots, p_N)$. The dynamics of the system is characterized by a Hamiltonian $H(q, p)$ and governed by Hamilton's equations

---

[1] The Hamiltonian does not depend on time explicitly.

© The Author(s), under exclusive license to Springer Nature Singapore Pte Ltd. 2021
K. Yoshida, *Yang–Baxter Deformation of 2D Non-Linear Sigma Models*,
SpringerBriefs in Mathematical Physics,
https://doi.org/10.1007/978-981-16-1703-4_1

$$\frac{dq^i}{dt} = \frac{\partial H(q, p)}{\partial p_i}, \qquad \frac{dp_i}{dt} = -\frac{\partial H(q, p)}{\partial q^i}. \tag{1.1}$$

If the system possesses N conserved charges I_i ($i = 1, \ldots, N$) that are independent and in involution, namely their Poisson brackets vanish,

$$\{I_i, I_j\}_P = 0 \qquad (^\forall i \text{ and } j = 1, \ldots, N),$$

then Hamilton's equations (1.1) can be solved in quadratures with the help of a canonical transformation from the original canonical pair (q^i, p_i) to a new one (θ^i, I_i), where θ^i is a conjugate coordinate to I_i. The components of the latter are called action-angle coordinates.

As a result, Hamilton's equations can be rewritten into

$$\frac{d\theta^i}{dt} = \frac{\partial \tilde{H}(\theta, I)}{\partial I_i}, \qquad \frac{dI_i}{dt} = -\frac{\partial \tilde{H}(\theta, I)}{\partial \theta^i}. \tag{1.2}$$

The second equation in (1.2) vanishes for all of i due to the conservation of I_i. Thus a new Hamiltonian $\tilde{H}(\theta, I)$ does not depend on θ^i. Then the right-hand side of the first equation in (1.2) also does not depend on θ^i and time t. So it can be easily solved as

$$\theta^i(t) = \Omega^i t + \theta^i(0), \tag{1.3}$$

where a time-independent constant Ω^i is now defined as

$$\Omega^i \equiv \frac{\partial \tilde{H}(\theta, I)}{\partial I_i}.$$

This is well known as the Arnold-Liouville theorem. Then the system is classically integrable in the sense of Liouville (complete integrability).

For complicated cases where the topology of the phase space is non-trivial, more careful discussion is necessary. This is beyond the scope of this book. For the detailed analysis, for example, see the standard textbooks [2, 3].

Beyond the Arnold-Liouville Theorem?

When the number of the degrees of freedom is *finite* like in classical mechanical systems, the notion of classical integrability is manifestly defined and it is familiar to even undergraduate students. But, what will happen in applying the same logic to *classical field theory*? In this case, the dynamical variables are $\phi(t, x^i)$, rather than $q^i(t)$, and they have continuum indices instead. At a glance, it seems likely that there may be no problem if there are an infinite number of conserved charges. However, it is not the case, because the Arnold-Liouville theorem cannot be applied any more in general, and even in (1+1)-dimensions. Hence the situation gets more complicated, and the concept of integrability itself needs to be reconsidered carefully.

In the following, we will be concerned with the classical integrability in (1+1)-dimensional classical field theory.

Lax Pair

A key to promoting the notion of the classical integrability from a mechanical system to a field theory in (1+1)-dimensions is Lax pair. To get used to the idea of Lax pair, it is worth revisiting here an integrable mechanical system.

When the mechanical system is classically integrable in the sense of Liouville, Hamilton's equation can be described by two matrices L and M which consist of functions on the phase space of the system. Namely, Hamilton's equations are reproduced by evaluating the first-order differential equation,

$$\frac{dL(t)}{dt} = [M(t), L(t)], \tag{1.4}$$

where $[M, L] \equiv ML - LM$ denotes the commutator of the matrices M and L. The pair (L, M) is called Lax pair.

A simple example[2] is a one-dimensional harmonic oscillator described by the following Hamiltonian,

$$H = \frac{1}{2}p^2 + \frac{1}{2}\omega^2 q^2, \tag{1.5}$$

where ω is the frequency and the mass is set to 1 for simplicity. Then Hamilton's equations are

$$\dot{q} = p, \quad \dot{p} = -\omega^2 q. \tag{1.6}$$

It is easy to check that the associated Lax pair is given by

$$L = \begin{pmatrix} p & \omega q \\ \omega q & -p \end{pmatrix}, \quad M = \begin{pmatrix} 0 & -\omega/2 \\ \omega/2 & 0 \end{pmatrix}. \tag{1.7}$$

As a matter of course, it is obviously waste of time to solve a harmonic oscillator by using the Lax pair description because we know much simpler methods to study its dynamics. However, in order to promote the notion of the classical integrability

---

[2]Although one usually would not try to solve the harmonic oscillator by using the Lax pair, it is still instructive to understand the notion of the integrability.

from mechanics to field theory, the Lax pair description is useful. In fact, the Lax decomposition of this type works well for classical field theory in (1+1)-dimensions as well, though some modifications are required.

As an example, let us consider the KdV equation,[3]

$$\frac{\partial u}{\partial t} = 6u\frac{\partial u}{\partial x} - \frac{\partial^3 u}{\partial x^3}, \tag{1.8}$$

where $u = u(t, x)$ is the dynamical variable. This differential equation can be rewritten into the Lax form,

$$L(t, x) = -\partial_x^2 + u, \qquad M(t, x) = -4\partial_x^3 + 6u\partial_x + 3\frac{\partial u}{\partial x}. \tag{1.9}$$

Here note that $L(t, x)$ is nothing but a Strum-Liouville operator.

It should be a good idea to employ the decomposition of the equations of motion into a Lax pair as a criterion of the classical integrability in classical field theory in (1+1)-dimensions. That is, when the field equation of motion can be decomposed into the Lax pair, the system is denoted to exhibit the *kinematical* integrability. This terminology is introduced, for example, in the textbook [13].

Note here that the kinematical integrability is different from the complete integrability in the sense of Liouville. Even for classical field theory in (1+1)-dimensions, it is possible to discuss the complete integrability by using the scattering data in the context of classical inverse scattering method and constructing the action-angle variables explicitly. The complete integrability has been shown, for example, for the KdV equation, sine-Gordon model and the Landau-Lifshitz model etc. (For the detail, see [13]).

However, it should be remarked that relativistic non-linear sigma models like principal chiral models and symmetric coset sigma models are kinematically integrable, but not complete integrable. This is because a non-ultra local term appears in the classical current algebra and causes ambiguity in integrating the current algebra [12]. For some approaches to improve this issue, see [7, 17]. At least so far, this is a long-standing problem in the field of classical integrable systems, and no satisfactory solution has been obtained so far.[4]

In the following, we are interested in the classical integrability of principal chiral models and symmetric coset sigma models. However, we will leave the issue of the non-locality as it is and concentrate on the kinematical integrability.

---

[3] For a comprehensive analysis of the KdV equation, see a very pedagogical textbook [6] by A. Das.

[4] As another perspective, the classical integrability may be discussed from the viewpoint of classical r-matrix. In the case with ambiguity of non-ultra local term, Maillet's r/s-formalism would be the best way at the moment.

1.2 Principal Chiral Model

In this section, let us consider a two-dimensional non-linear sigma model whose target space is given by a semi-simple Lie group G itself.[5] This model is called a 2D principal chiral model (PCM).[6]

For our notation, the two-dimensional base space is 2D Minkowski spacetime with the coordinates $x^\mu = (x^0, x^1) = (t, x)$ and the metric $\eta_{\mu\nu} = \mathrm{diag}(-1, +1)$. The group element $g \in G$ can be seen as a map from the base space to the target space G like $g(x)$. Namely, this $g(x)$ is a group element valued function. In the following, we will work with the fundamental representation of the Lie algebra $\mathfrak{g}$ of Lie group G. Hence it is assumed that the trace operation which appears below is defined for this fundamental representation.

1.2.1 Classical Action and Symmetry

The classical action of the G-principal chiral model in two dimensions is given by

$$S = -\frac{1}{2} \int d^2 x \, \eta^{\mu\nu} \mathrm{Tr}(J_\mu J_\nu) \,, \tag{1.10}$$

where J_μ is the left-invariant one-form (or equivalently the Maurer-Cartan one-form) defined as

$$J_\mu \equiv g^{-1} \partial_\mu g \tag{1.11}$$

in terms of the group element $g(x)$.

To be pedagogical, let us derive the classical equation of motion from the classical action (1.10). First of all, it is easy to see a variation of the left-invariant current as follows:

$$\delta J_\mu = \delta(g^{-1} \partial_\mu g) = \delta(g^{-1}) \partial_\mu g + g^{-1} \partial_\mu \delta g$$
$$= -g^{-1} \delta g \cdot J_\mu + g^{-1} \partial_\mu \delta g \,. \tag{1.12}$$

Note that

$$\delta(g^{-1}) = -g^{-1} \cdot \delta g \cdot g^{-1} \tag{1.13}$$

---

[5]We assume that G is a classical Lie group. The semi-simplicity is necessary to introduce the associated Killing form.

[6]In the following, we will work in the Lorentzian signature for the base space and the dimensionality (1+1) is abbreviated simply as 2D for brevity.

as shown by taking the variation of $g^{-1} \cdot g = 1$. Then, the variation of the classical action (1.10) is evaluated as

$$
\begin{aligned}
\delta S &= -\int d^2 x \, \eta^{\mu\nu} \text{Tr}(J_\mu \delta J_\nu) \\
&= -\int d^2 x \, \eta^{\mu\nu} \text{Tr}\Big(J_\mu J_\nu \cdot (-g^{-1}\delta g)\Big) - \int d^2 x \, \eta^{\mu\nu} \text{Tr}\Big(J_\mu \cdot (g^{-1}\partial_\nu \delta g)\Big).
\end{aligned}
\tag{1.14}
$$

By integrating by parts, the second term can be rewritten as

$$
\begin{aligned}
&\int d^2 x \, \eta^{\mu\nu} \text{Tr}\Big(J_\mu \cdot (g^{-1}\partial_\nu \delta g)\Big) \\
&= -\int d^2 x \, \eta^{\mu\nu} \text{Tr}\Big(\partial_\nu J_\mu \cdot (g^{-1}\delta g) + J_\mu \cdot (-J_\nu \cdot g^{-1})\delta g\Big),
\end{aligned}
\tag{1.15}
$$

where the surface term has been dropped off. By noting that the first term in (1.14) is canceled out by the second term in (1.15), the variation of the action is evaluated as

$$
\delta S = \int d^2 x \, \eta^{\mu\nu} \text{Tr}\Big(\partial_\nu J_\mu \cdot (g^{-1}\delta g)\Big),
\tag{1.16}
$$

and the classical equation of motion is given by

$$
\partial^\mu J_\mu = 0.
\tag{1.17}
$$

The classical action (1.10) enjoys the global symmetry $G \times G$. It is conventional to represent this symmetry as $G_L \times G_R$. The global G_L symmetry is defined as the left-multiplication,

$$
g(x) \longrightarrow g_L \cdot g(x) \qquad (^\forall g_L \in G_L),
\tag{1.18}
$$

where g_L does not depend on x^μ. This left symmetry (1.18) is manifest from the definition of the left-invariant one-form J. (This is the origin of the name of the *left-invariant* one-form.)

Similarly, the global right symmetry can also be introduced as the right multiplication defined as

$$
g(x) \longrightarrow g(x) \cdot g_R \qquad (^\forall g_R \in G_R),
\tag{1.19}
$$

where g_R does not depend on x^μ as well. To see that the classical action (1.10) is invariant under the global right symmetry, it is helpful to introduce the right-invariant one-form defined as

$$
I_\mu \equiv -\partial_\mu g \cdot g^{-1},
\tag{1.20}
$$

where the minus sign is taken for later convenience. This I_μ is manifestly invariant under the global right transformation (1.19). Then, due to the cyclic property of the trace, the classical action (1.10) can be rewritten as

$$S = -\frac{1}{2}\int d^2x \, \eta^{\mu\nu}\mathrm{Tr}(J_\mu J_\nu) = -\frac{1}{2}\int d^2x \, \eta^{\mu\nu}\mathrm{Tr}(I_\mu I_\nu)\,. \tag{1.21}$$

From the last expression, the global right symmetry (1.19) is manifest.

By following the standard Noether method, one can construct the conserved currents associated with the global G_L and G_R symmetries, respectively. As a result, the left-invariant one-form J_μ is a Noether current for the G_R symmetry, while the right-invariant one-form I_μ is for the G_L symmetry. Namely, the Noether currents j_μ^L and j_μ^R for $G_L \times G_R$, respectively, are given by

$$j_\mu^L \equiv I_\mu = -\partial_\mu g \cdot g^{-1} \qquad \text{(for } G_L)\,, \tag{1.22}$$
$$j_\mu^R \equiv J_\mu = g^{-1}\partial_\mu g \qquad \text{(for } G_R)\,. \tag{1.23}$$

The conservation laws for the currents are satisfied by the classical equation of motion for $g(x)$.

By definition, j_μ^L and j_μ^R satisfy the flatness condition, respectively,

$$\partial_\mu j_\nu^L - \partial_\nu j_\mu^L + [j_\mu^L, j_\nu^L] = 0\,, \tag{1.24}$$
$$\partial_\mu j_\nu^R - \partial_\nu j_\mu^R + [j_\mu^R, j_\nu^R] = 0\,. \tag{1.25}$$

This flatness condition will play a crucial role in ensuring the classical integrable structure. Note here that the above two expressions (1.24) and (1.25) are exactly the same because the extra minus has been included in the definition of the right-invariant one-form I_μ in (1.20). If this minus sign is not taken into account, then the sign in front of the commutator in (1.24) is flipped.

Finally, the principal chiral model enjoys a left-right symmetry called the g-parity,

$$\pi : g \longrightarrow g^{-1}\,. \tag{1.26}$$

For example, the left-invariant one-form J_μ is transformed to the right-invariant one-form I_μ like

$$J_\mu = g^{-1}\partial_\mu g \quad \longrightarrow \quad g\partial_\mu g^{-1} = g\cdot(-g^{-1}\partial_\mu g\cdot g^{-1})$$
$$= -\partial_\mu g\cdot g^{-1} = I_\mu\,. \tag{1.27}$$

The minus sign of the right-invariant one-form is useful here again.

1.2.2 Target-Space Geometry

Target Space of Non-linear Sigma Model

The target space of the principal chiral model is described in terms of the group element and hence it is not manifest to see the metric of the target space. To read off the relation between the target-space metric and the left-invariant one-form, it is useful to recall that the classical action of non-linear sigma model in a general set-up.

The classical action is given by

$$S_{\sigma M} = -\frac{1}{2}\int d^2x\, \eta^{\mu\nu} G_{MN}\partial_\mu X^M \partial_\nu X^N\,, \tag{1.28}$$

where X^M is a map from the world sheet with the coordinates $x^\mu = (\tau,\sigma)$ and the metric $\eta_{\mu\nu} = \mathrm{diag}(-1,1)$ to the target space with the coordinates X^M ($M = 1,\cdots,\dim M$ and the metric G_{MN}.

On the other hand, the classical action of the principal chiral model is given by

$$S = -\frac{1}{2}\int d^2x\, \eta^{\mu\nu}\mathrm{Tr}(J_\mu J_\nu)\,. \tag{1.29}$$

Supposing that the generators are normalized as

$$\mathrm{Tr}(T^a T^b) = \kappa_a \delta^{ab} \qquad (\kappa_a : \text{normalization const.})\,, \tag{1.30}$$

one can see the relation

$$\sum_{a=1}^{\dim G} \kappa_a J_\mu^a J_\nu^a = G_{MN}\partial_\mu X^M \partial_\nu X^N\,, \tag{1.31}$$

or equivalently,

$$\sum_{a=1}^{\dim G} \kappa_a J^a J^a = G_{MN} dX^M dX^N\,. \tag{1.32}$$

Thus the target-space metric G_{MN} is connected with the left-invariant one-form J.

Our goal here is to read off the metric G_{MN} by starting from the left-invariant one-form J with the use of the relation (1.32). For this purpose, it is still necessary to introduce a coordinate system for the target space via a parametrization of the group element g. In the following, we will see two examples, (1) 3D round sphere S^3 with $G = SU(2)$ and (2) 3D anti de Sitter space AdS$_3$ with $G = SL(2)$. As for the normalization of the metric G_{MN} for the convention utilized here, it is convenient to multiply $-1/2$ for the sphere and $1/2$ for the AdS. We will take this normalization hereafter.

(1) $G = SU(2)$: S^3

The first example is the case with $G = SU(2)$. The Lie group $SU(2)$ can be seen as a group manifold which is homeomorphic to S^3. Let us see how to derive the metric of round S^3 below.

In the first place, it is convenient to parametrize the group element $g(x)$ in terms of three angle variables $\phi(x)$, $\theta(x)$ and $\psi(x)$ as

$$g(x) = \exp\left[\phi(x)T^3\right]\exp\left[\theta(x)T^2\right]\exp\left[\psi(x)T^3\right], \tag{1.33}$$

where T^a $(a = 1, 2, 3)$ are the generators of Lie algebra $\mathfrak{su}(2)$,

$$[T^a, T^b] = \varepsilon_{abc}T^c, \tag{1.34}$$

and the structure constant ε_{abc} is the totally anti-symmetric tensor with $\varepsilon_{123} = +1$. Note here that the algebra (1.34) is written down in the convention that the generators are anti-Hermite. Hence the fundamental representation is realized by

$$T^a = -\frac{i}{2}\sigma^a, \tag{1.35}$$

with the standard Pauli matrices σ^a $(a = 1, 2, 3)$ defined as

$$\sigma^1 \equiv \begin{pmatrix} 0 & 1 \\ 1 & 0 \end{pmatrix}, \quad \sigma^2 \equiv \begin{pmatrix} 0 & -i \\ i & 0 \end{pmatrix}, \quad \sigma^3 \equiv \begin{pmatrix} 1 & 0 \\ 0 & -1 \end{pmatrix}, \tag{1.36}$$

so that the generators are normalized as

$$\text{Tr}(T^aT^b) = -\frac{1}{2}\delta^{ab}. \tag{1.37}$$

Next, the left-invariant one-form J_μ can be expanded as

$$\begin{aligned} J = J_\mu dx^\mu &= g^{-1}\partial_\mu g \cdot dx^\mu \quad (= g^{-1}dg) \\ &= \left(J_\mu^1 T^1 + J_\mu^2 T^2 + J_\mu^3 T^3\right) dx^\mu \\ &= J^1 T^1 + J^2 T^2 + J^3 T^3, \end{aligned} \tag{1.38}$$

and $J^a \equiv -2\text{Tr}(J\, T^a)$ $(a = 1, 2, 3)$ are expressed in terms of ϕ, θ and ψ as

$$J^1 = \sin\psi d\theta - \cos\psi \sin\theta d\phi \,,$$
$$J^2 = \cos\psi d\theta + \sin\psi \sin\theta d\phi \,,$$
$$J^3 = d\psi + \cos\theta d\phi \,. \tag{1.39}$$

Finally, the target-space metric is given by

$$ds^2 = -\frac{1}{2}\mathrm{Tr}(JJ) = \frac{1}{4}\sum_{a=1}^{3}(J^a)^2$$
$$= \frac{1}{4}\left[d\theta^2 + \sin^2\theta d\phi^2 + (d\psi + \cos\theta d\phi)^2 \right] . \tag{1.40}$$

This is nothing but the round S^3 metric. In this metric, S^3 is described as a $U(1)$-fibration over S^2. In this metric, only the left $SU(2)$ symmetry is manifest (while the right $SU(2)$ is implicitly realized). This result is consistent with the construction based on the left-invariant current, where only the left symmetry is manifest.

Note that S^3 can also be represented by a coset $SO(4)/SO(3)$ and then the homogeneity is obvious again. It is maximally symmetric and has a constant curvature.

(2) $G = SL(2)$: AdS$_3$

The second example is the case with $G = SL(2)$. Then, the group manifold $SL(2)$ itself is homeomorphic to AdS$_3$. In the following, we will derive the AdS$_3$ metric by taking an appropriate parametrization of the group element $g \in SL(2)$. The AdS space is also homogeneous, maximally symmetric and has a constant curvature.

1. The global AdS$_3$

First of all, the generators of $\mathfrak{sl}(2)$ are represented in terms of the Pauli matrices by

$$T^0 = \frac{i}{2}\sigma_1 \,, \qquad T^1 = \frac{1}{2}\sigma_2 \,, \qquad T^2 = \frac{1}{2}\sigma_3 \,. \tag{1.41}$$

The commutation relations are given by

$$[T^0, T^1] = -T^2 \,, \qquad [T^1, T^2] = T^0 \,, \qquad [T^2, T^0] = -T^1 \,. \tag{1.42}$$

It is useful to take the following parametrization of group element as

$$g = \exp\left[\tau(x)\, T^0\right] \exp\left[\sigma(x)\, T^2\right] \exp\left[u(x)\, T^1\right] \,, \tag{1.43}$$

where τ, σ and u are the target-space coordinates.

In the same way as in the case with $G = SU(2)$, by computing the square of J, the metric of the target space can be computed as

$$ds^2 = \frac{1}{2}\text{Tr}(JJ) = \frac{1}{4}\left[-\cosh^2\sigma\, d\tau^2 + d\sigma^2 + (du + \sinh\sigma\, d\tau)^2\right]. \quad (1.44)$$

This is the metric of AdS$_3$ in the global coordinates.

2. Poincare AdS$_3$

Next, let us examine another parametrization of group element,

$$g = \exp\left[2x^+ T^+\right]\exp\left[2(\log z)\, T^2\right]\exp\left[2x^- T^-\right], \quad (1.45)$$

where $x^\pm$ and $z\ (>0)$ are the target-space coordinates. Then $T^\pm$ are defined as

$$T^\pm \equiv i(T^1 \mp T^0), \quad (1.46)$$

and satisfy

$$[T^+, T^-] = 2T^2, \quad [T^2, T^\pm] = \pm T^\pm. \quad (1.47)$$

The numerical constants in the exponential factor have been adjusted so that a nice form of the AdS$_3$ metric should be obtained. The normalization of the generators are as follows:

$$\text{Tr}(T^+ T^-) = -1, \quad \text{Tr}(T^2 T^2) = \frac{1}{2}. \quad (1.48)$$

Then the left-invariant one-form J is expanded as

$$J = J^- T^+ + J^+ T^- + J^2 T^2, \quad (1.49)$$

and each of the components is expressed in terms of the target-space coordinates as

$$J^+ = -\text{Tr}(T^+ J) = 2dx^- + \frac{4x^-(2x^- dx^- - z\,dz)}{z^2},$$
$$J^- = -\text{Tr}(T^- J) = \frac{2dx^+}{z^2}, \quad (1.50)$$
$$J^2 = 2\text{Tr}(T^2 J) = \frac{2(-4x^- dx^+ + z\,dz)}{z^2}.$$

Similarly, the metric of the target space can be computed as

$$ds^2 = \frac{1}{2}\text{Tr}(JJ) = \frac{-4dx^+ dx^- + dz^2}{z^2}. \quad (1.51)$$

This is the metric of AdS$_3$ in the Poincare coordinates.

1.2.3 BIZZ Construction of Non-local Conserved Charges

One possible way to figure out the classical integrability of 2D principal chiral model
is to construct a Lax pair explicitly, as already explained in Sect. 1.1. But here, before
discussing the classical Lax pair, let us consider another way: the construction of an
infinite number of *non-local* conserved charges. As we will see later, it is closely
related to the existence of classical Lax pair. One reason to take this way in the first
place is that the construction of the conserved charges is really systematic, but there
is no general way to construct a Lax pair systematically. The construction of Lax
pair is usually based on an intuitive method and in general, case by case analysis is
unavoidable.

There are some methods to construct non-local conserved charges. We shall intro-
duce below an inductive method invented by Brezin, Itzykson, Zinn-Junstin and
Zuber (BIZZ) [5] (See also a comprehensive textbook [1]). This method is called the
BIZZ construction.

It is useful to introduce the light-cone coordinates for the base space

$$x^{\pm} = \frac{1}{2}(t \pm x), \qquad \partial_{\pm} = \partial_t \pm \partial_x. \tag{1.52}$$

Then the metric with the light-cone coordinates is

$$\eta_{+-} = \eta_{-+} = -2, \qquad \eta^{+-} = \eta^{-+} = -\frac{1}{2}. \tag{1.53}$$

The anti-symmetric tensor $\epsilon^{\mu\nu}$ on the base space is normalized as

$$\epsilon^{tx} = -\epsilon^{xt} = +1, \qquad \epsilon_{tx} = -\epsilon_{xt} = -1, \tag{1.54}$$

and in the light-cone coordinates,

$$\epsilon^{+-} = -\epsilon^{-+} = -\frac{1}{2}, \qquad \epsilon_{+-} = -\epsilon_{-+} = 2. \tag{1.55}$$

The convention of the light-cone coordinates (1.52) is not so popular, but the light-
cone components of the current J_{μ} is expressed as

$$J_{\pm} = J_t \pm J_x, \tag{1.56}$$

and it is rather useful in arguing the integrability below.

In the following, we will concentrate on the Noether current associated with the
left symmetry, j_{μ}^{L}. The same analysis is applicable to the right symmetry as well.

Let us take j_μ^L to be the zero-th current,

$$j_\mu^{(0)} \equiv j_\mu^L. \tag{1.57}$$

Now that the conservation law is written as

$$\partial_+ j_-^{(0)} + \partial_- j_+^{(0)} = 0, \tag{1.58}$$

$j_\mu^{(0)}$ can be expressed by using a Lie-algebra $\mathfrak{g}$-valued potential $\chi^{(0)}$ as follows:

$$j_\pm^{(0)} = \pm \partial_\pm \chi^{(0)}. \tag{1.59}$$

Then, by using $\chi^{(0)}$ and $j_\mu^{(0)}$, a new conserved current $j_\mu^{(1)}$ can be defined as

$$j_\mu^{(1)} \equiv \nabla_\mu^{(0)} \chi \equiv \partial_\mu \chi^{(0)} + \frac{1}{2}[j_\mu^{(0)}, \chi^{(0)}]. \tag{1.60}$$

Note here that the covariant derivative $\nabla_\mu^{(0)}$ is defined with $j_\mu^{(0)}/2$.

A remarkable point is that this new current $j_\mu^{(1)}$ is conserved due to the flatness condition of $j_\mu^{(0)}$. The conservation law can be shown as follows:

$$
\begin{aligned}
&\partial_+ j_-^{(1)} + \partial_- j_+^{(1)} \\
&= \partial_+ \left(\partial_- \chi^{(0)} + \frac{1}{2}[j_-^{(0)}, \chi^{(0)}] \right) + \partial_- \left(\partial_+ \chi^{(0)} + \frac{1}{2}[j_+^{(0)}, \chi^{(0)}] \right) \\
&= \partial_+ \left(-j_-^{(0)} + \frac{1}{2}[j_-^{(0)}, \chi^{(0)}] \right) + \partial_- \left(j_+^{(0)} + \frac{1}{2}[j_+^{(0)}, \chi^{(0)}] \right) \\
&= -\partial_+ j_-^{(0)} + \frac{1}{2}[\partial_+ j_-^{(0)}, \chi^{(0)}] + \frac{1}{2}[j_-^{(0)}, \partial_+ \chi^{(0)}] \\
&\quad + \partial_- j_+^{(0)} + \frac{1}{2}[\partial_- j_+^{(0)}, \chi^{(0)}] + \frac{1}{2}[j_+^{(0)}, \partial_- \chi^{(0)}] \\
&= -(\partial_+ j_-^{(0)} - \partial_- j_+^{(0)} + [j_+^{(0)}, j_-^{(0)}]) = 0. \tag{1.61}
\end{aligned}
$$

Thus $j_\mu^{(1)}$ is conserved due to the flatness condition of $j_\mu^{(0)}$.

Actually, the construction of $j_\mu^{(1)}$ is a bit special from the reasoning to be explained soon, but starting from this $j_\mu^{(1)}$, one can systematically construct an infinite number of conserved charges by induction.

From the conservation of $j_\mu^{(1)}$, a new potential $\chi^{(1)}$ can be introduced as

$$j_\pm^{(1)} = \pm \partial_\pm \chi^{(1)}. \tag{1.62}$$

Then a new conserved current $j_\mu^{(2)}$ is defined as

$$j_\mu^{(2)} \equiv \nabla_\mu \chi^{(1)} \equiv \partial_\mu \chi^{(1)} + [j_\mu^{(0)}, \chi^{(1)}]. \tag{1.63}$$

Here a covariant derivative ∇_μ with $j_\mu^{(0)}$ has been newly introduced. It should be remarked that this ∇_μ in (1.63) is different from the covariant derivative in (1.60) by the factor $1/2$ in front of the commutator. Then, as we will see below, an infinite number of conserved charges can be generated by induction by using the covariant derivative in (1.63), rather than the one in (1.60). This is the reason why the construction of $j_\mu^{(1)}$ is special. In addition, note also that the covariant derivative in (1.63) is defined with $j_\mu^{(0)}$, rather than $j_\mu^{(1)}$ and hence the conservation law of $j_\mu^{(2)}$ is satisfied thanks to the flatness condition of $j_\mu^{(0)}$. Actually, $j_\mu^{(1)}$ does not exhibit the flatness in general. In other words, the following argument is based only on the flatness of $j_\mu^{(0)}$ and the flatness of higher currents does not matter.

Let us show the conservation law of $j_\mu^{(2)}$. One can see that

$$\begin{aligned}
\partial_+ j_-^{(2)} + \partial_- j_+^{(2)} &= \partial_+ \nabla_- \chi^{(1)} + \partial_- \nabla_+ \chi^{(1)} \\
&= \partial_+ \left(\partial_- \chi^{(1)} + [j_-^{(0)}, \chi^{(1)}] \right) + \partial_- \left(\partial_+ \chi^{(1)} + [j_+^{(0)}, \chi^{(1)}] \right) \\
&= \partial_+ \left(-j_-^{(1)} + [j_-^{(0)}, \chi^{(1)}] \right) + \partial_- \left(j_+^{(1)} + [j_+^{(0)}, \chi^{(1)}] \right) \\
&= -\partial_+ j_-^{(1)} + [\partial_+ j_-^{(0)}, \chi^{(1)}] + [j_-^{(0)}, \partial_+ \chi^{(1)}] \\
&\quad + \partial_- j_+^{(1)} + [\partial_- j_+^{(0)}, \chi^{(1)}] + [j_+^{(0)}, \partial_- \chi^{(1)}] \\
&= -\partial_+ j_-^{(1)} + [j_-^{(0)}, j_+^{(1)}] + \partial_- j_+^{(1)} - [j_+^{(0)}, j_-^{(1)}]) \\
&= -\nabla_+ j_-^{(1)} + \nabla_- j_+^{(1)} \\
&= -\nabla_+ \nabla_- \chi^{(0)} + \nabla_- \nabla_+ \chi^{(0)} \\
&= -[\nabla_+, \nabla_-] \chi^{(0)} = 0. \tag{1.64}
\end{aligned}$$

Here at the last step the identity

$$[\nabla_\mu, \nabla_\nu] = 0, \tag{1.65}$$

which follows from the flatness condition of $j_\mu^{(0)}$, has been utilized. Thus, the mechanism to ensure the conservation law of $j_\mu^{(2)}$ is different from the one for $j_\mu^{(1)}$. Actually, the zero-curvature condition (1.65) is not satisfied by the covariant derivative in (1.60). In other words, the factor 2 is crucial for this zero-curvature condition.

Rigorously speaking, mathematical induction should be employed to show the existence of an infinite number of the charges. The case for $n = 2$ has already been constructed. Suppose that the n-th conserved current $j_\mu^{(n)}$ has been constructed by following our procedure. Namely, $j_\mu^{(n)}$ is conserved,

$$\partial_+ j_-^{(n)} + \partial_- j_+^{(n)} = 0, \tag{1.66}$$

and is written in terms of potentials $\chi^{(n)}$ and $\chi^{(n-1)}$ as

$$j_\pm^{(n)} = \pm\partial_\pm\chi^{(n)} = \nabla_\pm\chi^{(n-1)} . \tag{1.67}$$

Let us define the (n+1)-th current as

$$j_\mu^{(n+1)} \equiv \nabla_\mu\chi^{(n)} = \partial_\mu\chi^{(n)} + [j_\mu^{(0)}, \chi^{(n)}] . \tag{1.68}$$

Then the conservation law of $j_\mu^{(n+1)}$ can be shown as follows:

$$
\begin{aligned}
\partial_+ j_-^{(n+1)} + \partial_- j_+^{(n+1)} &= \partial_+ \nabla_-\chi^{(n)} + \partial_- \nabla_+\chi^{(n)} \\
&= \partial_+ \left(\partial_-\chi^{(n)} + [j_-^{(0)}, \chi^{(n)}]\right) + \partial_- \left(\partial_+\chi^{(n)} + [j_+^{(0)}, \chi^{(n)}]\right) \\
&= \partial_+ \left(-j_-^{(n)} + [j_-^{(0)}, \chi^{(n)}]\right) + \partial_- \left(j_+^{(n)} + [j_+^{(0)}, \chi^{(n)}]\right) \\
&= -\partial_+ j_-^{(n)} + [\partial_+ j_-^{(0)}, \chi^{(n)}] + [j_-^{(0)}, \partial_+\chi^{(n)}] \\
&\quad + \partial_- j_+^{(n)} + [\partial_- j_+^{(0)}, \chi^{(n)}] + [j_+^{(0)}, \partial_-\chi^{(n)}] \\
&= -\partial_+ j_-^{(n)} + [j_-^{(0)}, j_+^{(n)}] + \partial_- j_+^{(n)} - [j_+^{(0)}, j_-^{(n)}]) \\
&= -\nabla_+ j_-^{(n)} + \nabla_- j_+^{(n)} \\
&= -\nabla_+ \nabla_-\chi^{(n-1)} + \nabla_- \nabla_+\chi^{(n-1)} \\
&= -[\nabla_+, \nabla_-]\chi^{(n-1)} = 0 .
\end{aligned}
\tag{1.69}
$$

Here at the last step the identity (1.65) has been utilized again. Thus, $j_\mu^{(n+1)}$ is also conserved.

In summary, an infinite number of the conserved charges have been constructed and the associated Noether charges are defined as

$$Q^{(n)} = \int_{-\infty}^{\infty} dx\, j_t^{(n)}(t, x) . \tag{1.70}$$

Explicit Expressions of Non-local Charges

So far, we have not presented the explicit expressions of the potentials $\chi^{(n)}$. For simplicity, let us focus upon the first current $j_\mu^{(1)}$. It is easy to see that the equation in (1.59) is satisfied by

$$\chi^{(0)}(t, x) = \int_{-\infty}^{x} dy\, j_t^{(0)}(t, y) . \tag{1.71}$$

Indeed, by using the relations,

$$\partial_x \chi^{(0)}(t, x) = j_t^{(0)}(t, x),$$ (1.72)

$$\partial_t \chi^{(0)}(t, x) = \int_{-\infty}^{x} dy \, \partial_t j_t^{(0)}(t, y) = \int_{-\infty}^{x} dy \, \partial_y j_x^{(0)}(t, y)$$

$$= j_x^{(0)}(t, x),$$ (1.73)

one can easily check (1.59). Note that we have assumed here that the current $j_\mu^{(0)}$ vanishes at spatial infinity $x \to \pm\infty$. In fact, the integrability is sensitive to the choice of boundary conditions and there are some possible boundary conditions. Here we will work with the rapidly damping boundary condition so that all of the currents rapidly approach zero as $x \to \pm\infty$.

Then the current $j_\mu^{(1)}$ is expressed as

$$j_t^{(1)}(t, x) = j_x^{(0)}(t, x) + \frac{1}{2} \int_{-\infty}^{x} dy \, [j_t^{(0)}(t, x), j_t^{(0)}(t, y)],$$ (1.74)

$$j_x^{(1)}(t, x) = j_t^{(0)}(t, x) + \frac{1}{2} \int_{-\infty}^{x} dy \, [j_x^{(0)}(t, x), j_t^{(0)}(t, y)],$$ (1.75)

or equivalently,

$$j_\mu^{(1)}(t, x) = -\epsilon_{\mu\nu} j^{\nu(0)}(t, x) + \frac{1}{2} \int_{-\infty}^{x} dy \, [j_\mu^{(0)}(t, x), j_t^{(0)}(t, y)],$$ (1.76)

where $\epsilon^{\mu\nu}$ is the totally anti-symmetric tensor defined in (1.54). Thus the conserved charge $Q^{(1)}$ is given by

$$Q^{(1)} = \int_{-\infty}^{\infty} dx \, j_x^{(0)}(t, x) + \frac{1}{2} \int_{-\infty}^{\infty} dx \int_{-\infty}^{x} dy \, [j_t^{(0)}(t, x), j_t^{(0)}(t, y)].$$ (1.77)

So far, we have discussed $Q^{(1)}$ only, but the similar construction works for arbitrary $Q^{(n)}$ ($n \geq 2$) as well.

It should be remarked that the potential given in (1.71) is non-local. Hence it is obvious by construction that all of the currents $j_\mu^{(n)}$ and the associated conserved charges $Q^{(n)}$ constructed above are non-local for $n \geq 1$. The $n = 0$ case corresponds to the usual Noether current and charge.

In addition, the non-local charges constructed here do not commute with each other, and hence these are not in involution. It seems likely that this property is not consistent with the complete integrability in the sense of Liouville. However, the involution property is required for the *local* charges and not for the non-local ones. The non-local charges are still closely related to the classical integrability in the sense of the kinematical integrability (i.e., the existence of Lax pair), as we will see later.

1.2.4 Yangian Algebra

In the previous subsection, an infinite number of the conserved non-local charges have been constructed. It is worth seeing the algebra satisfied by these charges. To compute the algebra of the charges, it is necessary to prepare the classical Poisson algebra of the left-invariant one-form J_μ.

The classical Poisson algebra of the left-invariant current J_μ is given by

$$\{J_t^A(x), J_t^B(y)\}_P = f^{AB}{}_C J_t^C(x)\delta(x-y),\tag{1.78}$$

$$\{J_t^A(x), J_x^B(y)\}_P = f^{AB}{}_C J_x^C(x)\delta(x-y) + \delta^{AB}\partial_x\delta(x-y),\tag{1.79}$$

$$\{J_x^A(x), J_x^B(y)\}_P = 0,\tag{1.80}$$

where $\{\,,\,\}_P$ is the classical Poisson bracket in 2D principal chiral model and the brackets are evaluated at equal time t. The structure constant $f^{AB}{}_C$ defines the Lie algebra $\mathfrak{g}$ of the target-space Lie group G,

$$[T^A, T^B] = f^{AB}{}_C T^C.\tag{1.81}$$

Note that the commutation relation (1.79) contains a derivative of the delta function. This term is called the non-ultra local term. The appearance of this term is notorious because it gives rise to ambiguity in computing the algebra of non-local charges [12, 17]. As discussed in detail in Appendix, if it exists, the ordering of integrals in the definition of the non-local charges affects on the final result. For some argument to alleviate the non-ultra locality, see [7].

The existence of the non-ultra local term makes it quite difficult to discuss the complete integrability for principal chiral models and symmetric coset sigma models [12, 17], while in other classically integrable models such as sine-Gordon model and non-linear Schrödinger equation, the non-ultra local term does not appear in the current algebra, and the complete integrability can be shown by following the classical inverse scattering method (e.g., see [13]).

By following a possible prescription, the Poisson algebra of the non-local charges can be computed (up to the ambiguity). For simplicity, let us consider the case with $G = SU(2)$. Then the brackets for $Q^{(0)}$ and $Q^{(1)}$ are given by

$$\{Q^{(0)A}, Q^{(0)B}\}_P = \varepsilon^{AB}{}_C Q^{(0)C},\tag{1.82}$$

$$\{Q^{(0)A}, Q^{(1)B}\}_P = \varepsilon^{AB}{}_C Q^{(1)C},\tag{1.83}$$

$$\{Q^{(1)A}, Q^{(1)B}\}_P = \varepsilon^{AB}{}_C\left[Q^{(2)C} - \frac{1}{12}Q^{(0)C}Q^{(0)D}Q^{(0)}_D\right].\tag{1.84}$$

Here the first bracket (1.82) is just a sigma-model representation of the Lie algebra $\mathfrak{g} = \mathfrak{su}(2)$. The computation of the second bracket (1.83) is directly concerned with the ambiguity coming from the non-ultra local term. Here we took a certain prescription by following [17] (For the detail see Appendix). The third bracket is intrinsic to this algebra. The right-hand side is not linear with respect to the charges, and higher-order terms of the charges appear. This is not the usual Lie algebra but known as the Hopf algebra.[7] The coefficient $-1/12$ is specific to the $G = SU(2)$ case and it varies depending on the choice of G. The linear term in (1.84) is the next charge $Q^{(2)}$ and one can see that the label n of the charge describes the grade of the algebra.

The algebra presented in (1.82)–(1.84) is called the Yangian algebra, or simply Yangian [8, 9]. There are some ways to represent the Yangian and the above way is called Drinfeld's 1st realization. More rigorously speaking, we should mention about the Serre relation, which is necessary to preserve the tower structure of Yangian, but we will not discuss it here (For the detail, for example, see [16]).

1.2.5   Lax Pair and Monodromy Matrix

Lax Pair

In the first place, let us write down an expression of classical Lax pair:

$$\mathcal{L}_{\pm}(x; \lambda) \equiv \frac{J_{\pm}(x)}{1 \pm \lambda}, \tag{1.85}$$

where λ is a spectral parameter (complex constant).

It is easy to see that the flatness condition of the Lax pair

$$\partial_+ \mathcal{L}_- - \partial_- \mathcal{L}_+ + [\mathcal{L}_+, \mathcal{L}_-] = 0 \tag{1.86}$$

is equivalent to the equation of motion and the flatness condition of J_μ. Indeed, the left-hand side of (1.86) is evaluated as

$$
\begin{aligned}
& \partial_+ \mathcal{L}_- - \partial_- \mathcal{L}_+ + [\mathcal{L}_+, \mathcal{L}_-] \\
&= \frac{1}{1-\lambda} \partial_+ J_- - \frac{1}{1+\lambda} \partial_- J_+ + \left[\frac{J_+}{1+\lambda}, \frac{J_-}{1-\lambda} \right] \\
&= \frac{1}{1-\lambda^2} \Big((1+\lambda)\partial_+ J_- - (1-\lambda)\partial_- J_+ + [J_+, J_-] \Big) \\
&= \frac{1}{1-\lambda^2} \Big(\partial_+ J_- - \partial_- J_+ + [J_+, J_-] + \lambda(\partial_+ J_- + \partial_- J_+) \Big) \\
&= 0.
\end{aligned}
\tag{1.87}
$$

---

[7] We should check the defining relations of the Hopf algebra, but we will not do that. Those who are interested in them should refer to reviews on Yangian (for example, [4, 16]).

Here the flatness condition and equation of motion for J_μ have been utilized at the last equality.

The Lax pair in (1.85) is expressed in terms of the light-cone coordinates. It is easy to see that the covariant form of the Lax pair is given by

$$\mathcal{L}_\mu(x; \lambda) = \frac{1}{1 - \lambda^2} \left[J_\mu + \lambda \epsilon_{\mu\nu} J^\nu \right]. \tag{1.88}$$

Then the flatness condition can be written in the covariant way:

$$\partial_\mu \mathcal{L}_\nu - \partial_\nu \mathcal{L}_\mu + [\mathcal{L}_\mu, \mathcal{L}_\nu] = 0 \tag{1.89}$$

and it is equivalent to

$$[\partial_\mu + \mathcal{L}_\mu, \partial_\nu + \mathcal{L}_\nu] = 0. \tag{1.90}$$

It is worth noting that the Lax pair presented here does not look like the one introduced in Sect. 1.1. But these two expressions are closely related. We need to prepare some technical knowledge to explain the relation and it would be beyond the level of this book. For the detail, for example, see Sect. 3.7 of the standard text book [3].

Monodromy Matrix

The next is to introduce a generating function to provide an infinite number of the non-local charges constructed by the BIZZ construction. This generating function is the so-called monodromy matrix.

By using the spatial component of the classical Lax pair, the monodromy matrix $M(t; \lambda)$ can be defined as

$$
\begin{aligned}
M(t; \lambda) &= P \exp\left[-\int_{-\infty}^{\infty} dx\, \mathcal{L}_x(t, x; \lambda) \right] \\
&= P \exp\left[-\frac{1}{1 - \lambda^2} \int_{-\infty}^{\infty} dx\, (J_x - \lambda J_t) \right],
\end{aligned} \tag{1.91}
$$

where the symbol P denotes the path ordering of integrals as usual:

$$
\begin{aligned}
&\frac{1}{n!} \int_{x_0}^{x} dx_1 \cdots \int_{x_0}^{x} dx_n\, P\{ f(x_1) \cdots f(x_n) \} \\
&= \int_{x_0}^{x} dx_1 \int_{x_0}^{x_1} dx_2 \cdots \int_{x_0}^{x_{n-1}} dx_n\, f(x_1) \cdots f(x_n).
\end{aligned} \tag{1.92}
$$

For example, in the $n = 2$ case, P acts like

$$
\int_{x_0}^{x} dx_1 \int_{x_0}^{x} dx_2 \, P\{f(x_1)f(x_2)\}
$$

$$
= \int_{x_0}^{x} dx_1 \int_{x_0}^{x} dx_2 \left[\theta(x_1 - x_2)f(x_1)f(x_2) + \theta(x_2 - x_1)f(x_2)f(x_1)\right]
$$

$$
= \int_{x_0}^{x} dx_1 \int_{x_0}^{x} dx_2 \, \theta(x_1 - x_2)f(x_1)f(x_2) + \int_{x_0}^{x} dx_2 \int_{x_0}^{x} dx_1 \, \theta(x_1 - x_2)f(x_1)f(x_2)
$$

$$
= 2 \int_{x_0}^{x} dx_1 \int_{x_0}^{x_1} dx_2 \, f(x_1)f(x_2) , \tag{1.93}
$$

where $\theta(x)$ is the step function defined as

$$
\theta(x) = \begin{cases} 1 & (x > 0) \\ 0 & (x < 0) \end{cases} . \tag{1.94}
$$

For later convenience, it is also instructive to rewrite the second line in (1.93) as

$$
\int_{x_0}^{x} dx_1 \int_{x_0}^{x} dx_2 \, P\{f(x_1)f(x_2)\}
$$

$$
= \int_{x_0}^{x} dx_1 \int_{x_0}^{x} dx_2 \left[\theta(x_1 - x_2)f(x_1)f(x_2) + (1 - \theta(x_1 - x_2))f(x_2)f(x_1)\right]
$$

$$
= \int_{x_0}^{x} dx_2 \, f(x_2) \int_{x_0}^{x} dx_1 \, f(x_1) + \int_{x_0}^{x} dx_1 \int_{x_0}^{x_1} dx_2 \, [f(x_1), f(x_2)] , \tag{1.95}
$$

where we have used the relation

$$
\theta(x_1 - x_2) + \theta(x_2 - x_1) = 1 .
$$

A remarkable property of the monodromy matrix (1.91) is that it is conserved:

$$
\frac{\partial}{\partial t} M(t; \lambda) = 0 . \tag{1.96}
$$

The proof of it is the following:

$$\frac{\partial}{\partial t} M(t; \lambda)$$

$$= \int_{-\infty}^{\infty} dx\, P \exp\left[-\int_{x}^{\infty} dy\, \mathcal{L}_x(t, y; \lambda)\right] (\partial_t \mathcal{L}_x)\, P \exp\left[-\int_{-\infty}^{x} dy\, \mathcal{L}_x(t, y; \lambda)\right]$$

$$= \int_{-\infty}^{\infty} dx\, P \exp\left[-\int_{x}^{\infty} dy\, \mathcal{L}_x(t, y; \lambda)\right] \left(\partial_x \mathcal{L}_t - [\mathcal{L}_t, \mathcal{L}_x]\right)$$

$$\times P \exp\left[-\int_{-\infty}^{x} dy\, \mathcal{L}_x(t, y; \lambda)\right]$$

$$= \int_{-\infty}^{\infty} dx\, \partial_x \left(P \exp\left[-\int_{x}^{\infty} dy\, \mathcal{L}_x(t, y; \lambda)\right] (\mathcal{L}_t)\, P \exp\left[-\int_{-\infty}^{x} dy\, \mathcal{L}_x(t, y; \lambda)\right]\right)$$

$$= \mathcal{L}_t(x = \infty) M(t; \lambda) - M(t; \lambda) \mathcal{L}_t(x = -\infty)$$

$$= 0. \tag{1.97}$$

Here at the last step, we have used the rapidly damping boundary condition. Thus the manodromy matrix (1.91) is conserved due to the flatness condition of Lax pair (1.89) and the boundary condition.

Hence after expanding this monodromy matrix around a certain value of λ, each term gives rise to conserved charges. So it is quite easy to get an infinite number of the conserved quantities. The expressions of the charges depend on the expansion points. As shown in the following, $\lambda = \infty$ corresponds to the non-local charges constructed by the BIZZ construction.

Let us expand formally the monodromy matrix (1.91) around $\lambda = \infty$. With the notation $\tilde{\lambda} \equiv 1/\lambda$, the monodromy matrix is expanded as

$$M(t; \lambda)$$

$$= 1 - \tilde{\lambda} \int_{-\infty}^{\infty} dx\, J_t + \tilde{\lambda}^2 \left[\int_{-\infty}^{\infty} dx\, J_x + \frac{1}{2} \int_{-\infty}^{\infty} dx \int_{-\infty}^{\infty} dy\, P\{J_t(x) J_t(y)\}\right] + O(\tilde{\lambda}^3)$$

$$= 1 - \tilde{\lambda} Q^{(0)} + \frac{1}{2} \tilde{\lambda}^2 \left(Q^{(1)} + Q^{(0)} Q^{(0)}\right) + O(\tilde{\lambda}^3). \tag{1.98}$$

The second term corresponds to the zero-th Noether charge and the third term contains the first non-local charge and the square of the zero-th charges. Here the first a few terms have been explicitly computed, but it is possible in principle to compute higher-order terms systematically. Eventually, the power of each term in the expansion corresponds to the grade of Yangian.

1.3 Symmetric Coset Sigma Model

In this section, let us consider a 2D non-linear sigma model whose target space is given by a symmetric coset $M = G/H$, where G is a semi-simple Lie group and H is a subgroup of G. This model is simply called a symmetric coset sigma model. This sigma model also exhibits the classical integrability in the sense of the kinematical integrability, as shown below. It is remarkable that the integrability is universally ensured for any symmetric coset.

First of all, we will introduce the definition of the symmetric coset. Then the classical action of the symmetric coset sigma model is introduced. Due to the coset structure, the left G symmetry still survives as a global symmetry, while the right H symmetry is gauged.

Some general properties of the symmetric coset sigma model will be described. In particular, it will be shown that the symmetric coset structure gives rise to the flat conserved current universally. This is quite a significant property of the symmetric coset sigma model, because the existence of the flat conserved current leads to the classical Lax pair, Yangian and monodromy matrix etc. and the classical integrability in the kinematical sense is automatically ensured due to the symmetric coset structure.

1.3.1 Symmetric Coset

Let us introduce the definition of symmetric coset below.

Given a coset $M = G/H$, the Lie algebras of G and H are represented by $\mathfrak{g}$ and $\mathfrak{h}$, respectively. Then the Lie algebra $\mathfrak{g}$ can be decomposed into the direct product form

$$\mathfrak{g} = \mathfrak{h} \oplus \mathfrak{m}. \tag{1.99}$$

Here the product $\oplus$ is defined as the one for vector spaces, not Lie algebras.

If the coset $M = G/H$ satisfies the following properties:

$$[\mathfrak{h}, \mathfrak{h}] \subset \mathfrak{h}, \quad [\mathfrak{h}, \mathfrak{m}] \subset \mathfrak{m}, \quad [\mathfrak{m}, \mathfrak{m}] \subset \mathfrak{h}, \tag{1.100}$$

then the coset is called a symmetric coset. By assigning the grade as 0 for $\mathfrak{h}$ and 1 for $\mathfrak{m}$, this coset enjoys the $\mathbb{Z}_2$-grading.

As a simple example, let us consider $M = SU(2)/U(1)$, namely $G = SU(2)$ and $H = U(1)$. The Lie algebra of $SU(2)$, $\mathfrak{su}(2)$ is spanned by the set of generators: $\{T^3, T^\pm\}$. Then the Lie algebra of H, $\mathfrak{h}$ is represented by $\{T^3\}$ and the associated vector space $\mathfrak{m}$ is $\{T^+, T^-\}$. The commutation relations of the generators are given by

$$[T^3, T^3] = 0, \quad [T^3, T^\pm] = \mp i T^\pm, \quad [T^+, T^-] = 2i T^3, \tag{1.101}$$

where the $\mathfrak{su}(2)$ generators are given in (1.36) and we have introduced

$$T^\pm \equiv i(T^1 \pm i T^2). \tag{1.102}$$

Hence the matrix representation is now given by

$$T^3 = -\frac{i}{2}\begin{pmatrix} 1 & 0 \\ 0 & -1 \end{pmatrix}, \qquad T^+ = \begin{pmatrix} 0 & 1 \\ 0 & 0 \end{pmatrix}, \qquad T^- = \begin{pmatrix} 0 & 0 \\ 1 & 0 \end{pmatrix}. \qquad (1.103)$$

By assigning the grade 0 for T^3 and 1 for $T^\pm$, the $\mathbb{Z}_2$-grading is manifestly realized.

So far, we have explained purely a mathematical nature of symmetric coset. We will see soon later that this $\mathbb{Z}_2$-grading plays an important role so as to ensure the classical integrability of the 2D symmetric coset sigma model.

Coset Projector

For later purpose, it is convenient to introduce a coset projector $P : \mathfrak{g} \to \mathfrak{m}$ in order to realize the symmetric coset structure. This projection can be defined as

$$P(X) \equiv \sum_{a=1}^{\dim(\mathfrak{m})} \frac{\mathrm{Tr}(T^a X)}{\mathrm{Tr}(T^a T^a)} T^a \qquad (\forall X \in \mathfrak{g}). \qquad (1.104)$$

1.3.2 Classical Action and Symmetry

By using the coset projector P defined in (1.104), the classical action of the 2D symmetric coset sigma model with $M = G/H$ can be described as

$$\begin{aligned} S_{\mathrm{symm}} &= -\frac{1}{2}\int d^2x\, \eta^{\mu\nu}\mathrm{Tr}\left(J_\mu P(J_\nu)\right) \\ &\equiv -\frac{1}{2}\int d^2x\, \mathrm{Tr}\left(k_\mu k^\mu\right), \end{aligned} \qquad (1.105)$$

where the left-invariant one-form J_μ is defined in (1.11) and we have introduced a new notation

$$k_\mu \equiv P(J_\mu). \qquad (1.106)$$

Let us see the symmetry of the classical action (1.105). By definition, the left *global* G-symmetry

$$g \longrightarrow g_{\mathrm{L}} \cdot g \qquad (1.107)$$

is manifest. In comparison to the principal chiral model, the right symmetry is realized as a local symmetry

$$g \longrightarrow g \cdot h \quad (h \in H), \tag{1.108}$$

where h depends on the base-space coordinates x^μ. Then the left-invariant one-form J_μ is transformed as

$$
\begin{aligned}
J'_\mu &= (g \cdot h)^{-1} \partial_\mu (g \cdot h) \\
&= h^{-1} g^{-1} (\partial_\mu g \cdot h + g \cdot \partial_\mu h) \\
&= h^{-1} J_\mu h + h^{-1} \partial_\mu h .
\end{aligned} \tag{1.109}
$$

Thus the right *local H*-symmetry is regarded as a gauge transformation. Based on this fact, it should be useful to introduce a covariant derivative for this gauge symmetry. Note that the left-invariant one-form J_μ can be decomposed as

$$J_\mu = g^{-1} \partial_\mu g = P(J_\mu) + A_\mu = k_\mu + A_\mu , \tag{1.110}$$

where A_μ takes the values in the Lie algebra $\mathfrak{h}$ of H and k_μ belongs to $\mathfrak{m}$ by construction. Let us introduce a covariant derivative D_μ as

$$D_\mu g \equiv \partial_\mu g - g A_\mu . \tag{1.111}$$

By recalling the gauge transformation laws under $g \to g' = g \cdot h$,

$$
\begin{aligned}
A_\mu &\to A'_\mu = h^{-1} \cdot A_\mu \cdot h + h^{-1} \partial_\mu h , \\
D_\mu g &\to D'_\mu g' = D_\mu g \cdot h , \\
k_\mu &\to k'_\mu = h^{-1} \cdot k_\mu \cdot h ,
\end{aligned} \tag{1.112}
$$

the covariant derivatives of $D_\mu g$ and k_μ can be defined as, respectively,

$$
\begin{aligned}
D_\mu D_\nu g &\equiv \partial_\mu D_\nu g - (D_\nu g) \cdot A_\mu , \\
D_\mu k_\nu &\equiv \partial_\mu k_\nu + [A_\mu, k_\nu] .
\end{aligned} \tag{1.113}
$$

These expressions will appear in describing the equation of motion.

Then, from (1.110), one finds that

$$D_\mu g = g k_\mu \quad \text{or equivalently} \quad k_\mu = g^{-1} D_\mu g . \tag{1.114}$$

Thus the classical action may be written with the covariant derivative as

$$S_{\text{symm}} = -\frac{1}{2} \int d^2x \, \text{Tr} \left(g^{-1} D_\mu g \cdot g^{-1} D^\mu g \right) . \tag{1.115}$$

This expression can also be obtained by gauging the right symmetry in the principal chiral model.

The equation of motion can be derived from (1.115) and is expressed as

$$D^\mu D_\mu g - D_\mu g \cdot g^{-1} \cdot D^\mu g = 0, \tag{1.116}$$

or equivalently,

$$D^\mu k_\mu = 0. \tag{1.117}$$

By using (1.113), the last expression can be further rewritten into

$$\partial_\mu k^\mu + [A_\mu, k^\mu] = 0. \tag{1.118}$$

Note here that the equation of motion is a quantity with grade 1, while the covariant derivative D_μ is with grade 0.

The Noether current associated with the left G-symmetry can be derived as

$$j_\mu \equiv -2D_\mu g \cdot g^{-1} = -2g k_\mu g^{-1}. \tag{1.119}$$

Here the factor 2 has been introduced for later convenience. It is easy to see the conservation law of j_μ,

$$\partial^\mu j_\mu = 0, \tag{1.120}$$

by using the expression (1.116) of the equation of motion.

In addition, the Noether current j_μ also satisfies the flatness condition[8]

$$\partial_\mu j_\nu - \partial_\nu j_\mu + [j_\mu, j_\nu] = 0. \tag{1.121}$$

This means that there always exists a flat conserved current for a symmetric coset sigma model. Therefore, the analysis on the left G-symmetry in the principal chiral model can be applied in the same way. So, for the left G-symmetry, it is straightforward to construct Yangian generators, Lax pair, monodromy matrix etc. For later discussion, it is helpful to write down the expression of Lax pair explicitly:

$$\mathcal{L}_\pm(x; \lambda) \equiv \frac{j_\pm}{1 \pm \lambda} = \frac{-2g k_\pm g^{-1}}{1 \pm \lambda}. \tag{1.122}$$

Since the proof of the flatness condition (1.121) is a bit intricate, we will show it in detail in the next subsection.

---

[8]To get the same expression of the flatness condition, the factor 2 has been introduced in (1.119). If not, the factor 2 is multiplied in front of the commutator.

1.3.3   The Proof of the Flatness Condition

Let us here prove the flatness condition of the Noether current j_μ. First of all, we shall evaluate the derivative of j_μ as follows:

$$
\begin{aligned}
\partial_\mu j_\nu &= -2\partial_\mu(gk_\nu g^{-1}) \\
&= 2g\left[-\partial_\mu k_\nu + [k_\nu, g^{-1}\partial_\mu g]\right]g^{-1} \\
&= 2g\left(-D_\mu k_\nu - [k_\mu, k_\nu]\right)g^{-1}.
\end{aligned}
\tag{1.123}
$$

Similarly, one can derive

$$
-\partial_\nu j_\mu = 2g\left(D_\nu k_\mu + [k_\nu, k_\mu]\right)g^{-1}.
\tag{1.124}
$$

Finally, the last piece is evaluated as

$$
+[j_\mu, j_\nu] = 4g[k_\mu, k_\nu]g^{-1}.
\tag{1.125}
$$

Thus one obtains

$$
\partial_\mu j_\nu - \partial_\nu j_\mu + [j_\mu, j_\nu] = 2g\left(-D_\mu k_\nu + D_\nu k_\mu\right)g^{-1}.
\tag{1.126}
$$

Then the remaining task is to show that

$$
D_\mu k_\nu - D_\nu k_\mu = 0.
\tag{1.127}
$$

The derivation of (1.127) is a bit tricky. One has to start from the identity satisfied by the left-invariant one-form $J_\mu = g^{-1}\partial_\mu g$. First,

$$
\partial_\mu(g^{-1}\partial_\nu g) = -g^{-1}\partial_\mu g \cdot g^{-1}\partial_\nu g + g^{-1}\partial_\mu\partial_\nu g.
\tag{1.128}
$$

Similarly,

$$
-\partial_\nu(g^{-1}\partial_\mu g) = g^{-1}\partial_\nu g \cdot g^{-1}\partial_\mu g - g^{-1}\partial_\nu\partial_\mu g.
\tag{1.129}
$$

Therefore, by combining (1.128) and (1.129), one can show that

$$
\partial_\mu(g^{-1}\partial_\nu g) - \partial_\nu(g^{-1}\partial_\mu g) = -[g^{-1}\partial_\mu g, g^{-1}\partial_\nu g].
\tag{1.130}
$$

and thus the left-invariant one-form identically satisfies the flatness condition,

$$
\partial_\mu(g^{-1}\partial_\nu g) - \partial_\nu(g^{-1}\partial_\mu g) + [g^{-1}\partial_\mu g, g^{-1}\partial_\nu g] = 0.
\tag{1.131}
$$

Note that this is the flatness condition for the left-invariant current and not for the Noether current, though the former leads to the latter, as we will see below.

Here, by noting that $g^{-1}\partial_\mu g = k_\mu + A_\mu$, let us act the projection operator P on the expression in (1.131). Then one obtains

$$\partial_\mu k_\nu - \partial_\nu k_\mu + [k_\mu, A_\nu] + [A_\mu, k_\nu] = 0, \tag{1.132}$$

due to the symmetric coset structure. Namely,

$$[k_\mu, k_\nu] \in \mathfrak{h}, \quad [k_\mu, A_\nu] \in \mathfrak{m}, \quad [A_\mu, A_\nu] \in \mathfrak{h}. \tag{1.133}$$

This expression is equivalent to (1.127). Consequently, the Noether current j_μ satisfies the flatness condition.

1.3.4 Example: Coset Construction of S^2

Here let us compute the metric of S^2 based on its symmetric coset structure. The projection operator is given by

$$P(X) = \frac{\text{Tr}(T^- X)}{\text{Tr}(T^+ T^-)} T^+ + \frac{\text{Tr}(T^+ X)}{\text{Tr}(T^+ T^-)} T^-, \tag{1.134}$$

where the generators $T^\pm$ and T^3 are given in (1.103).

Then, by taking a parametrization of group element g as[9]

$$g = \exp\left[-\phi(x)\frac{i}{2}(T^+ + T^-)\right] \exp\left[\left(\theta(x) - \frac{\pi}{2}\right)\frac{1}{2}(T^+ - T^-)\right], \tag{1.135}$$

the target-space metric is given by

$$ds^2 = -2\text{Tr}(J P(J)) = d\theta^2 + \sin^2\theta\, d\phi^2. \tag{1.136}$$

Note here that the expression of the metric has followed from the sigma-model action (1.105).

---

[9]This is essentially the same as (1.33), up to the shift of θ by $\pi/2$, after setting $\psi = 0$.

1.3.5 Another Form of Lax Pair

In the previous subsection, a Lax pair has been presented by constructing a flat conserved current. Here, we shall introduce another form of Lax pair, in which the grading property is manifest.

Preparation

Recall that the left-invariant one-form J can be decomposed as

$$J = A + k \equiv J^{(0)} + J^{(1)}. \tag{1.137}$$

Here, to show the grading property explicitly, we have redefined as

$$J^{(0)} \equiv A, \quad J^{(1)} \equiv k.$$

Then the equation of motion (1.118) is expressed as

$$\partial_\mu J^{\mu(1)} + [J_\mu^{(0)}, J^{\mu(1)}] = 0. \tag{1.138}$$

In terms of the light-cone coordinates, this can be rewritten as

$$\partial_+ J_-^{(1)} + \partial_- J_+^{(1)} + [J_+^{(0)}, J_-^{(1)}] - [J_-^{(1)}, J_+^{(0)}] = 0. \tag{1.139}$$

Furthermore, the flatness condition of J

$$\partial_+ J_- - \partial_- J_+ + [J_+, J_-] = 0$$

can also be decomposed by the grading like

$$\partial_+ J_-^{(0)} - \partial_- J_+^{(0)} + [J_+^{(0)}, J_-^{(0)}] + [J_+^{(1)}, J_-^{(1)}] = 0, \tag{1.140}$$
$$\partial_+ J_-^{(1)} - \partial_- J_+^{(1)} + [J_+^{(0)}, J_-^{(1)}] + [J_+^{(1)}, J_-^{(0)}] = 0. \tag{1.141}$$

By combining (1.139) and (1.141), one can show that

$$\partial_+ J_-^{(1)} + [J_+^{(0)}, J_-^{(1)}] = 0, \tag{1.142}$$
$$\partial_- J_+^{(1)} - [J_+^{(1)}, J_-^{(0)}] = 0. \tag{1.143}$$

Another Expression

Another expression of the Lax pair is given by

$$\tilde{\mathcal{L}}_\pm(x; \tilde{\lambda}) \equiv J_\pm^{(0)} + \tilde{\lambda}^{\pm 1} J_\pm^{(1)}, \tag{1.144}$$

where $\tilde{\lambda} \in \mathbb{C}$ is a spectral parameter but different from λ in (1.122). It is easy to see the flatness condition

$$\partial_+ \tilde{\mathcal{L}}_- - \partial_- \tilde{\mathcal{L}}_+ + [\tilde{\mathcal{L}}_+, \tilde{\mathcal{L}}_-] = 0 \tag{1.145}$$

is equivalent to (1.140), (1.142) and (1.143).

It is instructive to remark some points. The Lax pair (1.144) is constructed by the left-invariant one-form J itself, which satisfies the flatness condition by definition but is not conserved. Then, while the previous Lax pair $\mathcal{L}$ in (1.122) has two poles $\lambda = \pm 1$, the poles of the Lax pair $\tilde{\mathcal{L}}$ in (1.144) are located on $\tilde{\lambda} = 0$ and ∞. However, λ and $\tilde{\lambda}$ are related through the relation

$$\tilde{\lambda} = \frac{\lambda - 1}{\lambda + 1}, \tag{1.146}$$

and these two Lax pairs are gauge-equivalent as shown in the last paragraph.

We will use the expression of Lax pair (1.144) in the context of Yang-Baxter deformation in the next chapter. In this expression the grading structure is manifest and it is easy to generalize it to higher-grading cases. In particular, it is convenient for the $\mathbb{Z}_4$-grading of $SU(2, 2|4)$, which is relevant to the standard example of AdS/CFT.

Gauge Equivalence

Let us show that the two expressions of the Lax pair (1.122) and (1.144) are equivalent under a gauge transformation with the relation (1.146).

Recalling that $g^{-1} dg = J^{(0)} + J^{(1)}$, we can rewrite the Lax pair (1.144) as

$$\tilde{\mathcal{L}}_+ = (\tilde{\lambda} - 1) J_+^{(1)} + g^{-1} \partial_+ g, \qquad \tilde{\mathcal{L}}_- = (\tilde{\lambda}^{-1} - 1) J_-^{(1)} + g^{-1} \partial_+ g \tag{1.147}$$

by removing $J^{(0)}$.

It is important to remark that the relation (1.145) is invariant under a gauge transformation

$$\tilde{\mathcal{L}}_\pm \longrightarrow \tilde{\mathcal{L}}'_\pm = g \tilde{\mathcal{L}}_\pm g^{-1} - \partial_\pm g \cdot g^{-1}. \tag{1.148}$$

That is, there is a gauge degree of freedom in describing the Lax pair and its expression is not unique in general. By rewriting the transformation law (1.148) as

$$\tilde{\mathcal{L}}_\pm = g^{-1} \tilde{\mathcal{L}}'_\pm g + g^{-1} \partial_\pm g, \tag{1.149}$$

we can read off $\tilde{\mathcal{L}}'$ as

$$\tilde{\mathcal{L}}'_{\pm} \equiv (\tilde{\lambda}^{\pm 1} - 1) g J^{(1)}_{\pm} g^{-1}. \tag{1.150}$$

By noting that $J^{(1)} = k$ and comparing (1.122) and (1.150), the following two relations should be satisfied

$$\tilde{\lambda}^{\pm 1} - 1 = \frac{-2}{1 \pm \lambda}. \tag{1.151}$$

However, both of them are nothing but the relation (1.146)!

Thus, we have shown that the two expressions of the Lax pair (1.122) and (1.144) are equivalent under a gauge transformation with the relation (1.146).

Appendix

In computing the charge algebra, an ambiguity for the ordering of the integrals arises from the delta-prime term. This ambiguity will be described explicitly below.

Let us focus upon the first two charges $Q^{(0)}$ and $Q^{(1)}$ and regularize the integration range as

$$Q^{(0) A} = \lim_{X_1, X_2 \to \infty} \int_{-X_1}^{X_2} dx \, J_t^A(x),$$

$$Q^{(1) A} = \lim_{Y_1, Y_2 \to \infty} \left[\int_{-Y_1}^{Y_2} dx \, J_x^A(x) + \frac{1}{2} f_{BC}{}^A \int_{-Y_1}^{Y_2} dx \int_{-Y_1}^{Y_2} dy \, \theta(x - y) J_t^B(x) J_t^C(y) \right],$$

where X_1, X_2, Y_1 and Y_2 are assumed to be positive real constant.

There is no problem for the Poisson bracket of two $Q^{(0)}$'s. The problem occurs in computing the Poisson bracket of $Q^{(0)}$ and $Q^{(1)}$ as follows:

$$\begin{aligned} \{Q^{(0)A}, Q^{(1)B}\}_{\rm P} &= \lim_{X_1, X_2 \to \infty} \lim_{Y_1, Y_2 \to \infty} \int_{-X_1}^{X_2} dx \int_{-Y_1}^{Y_2} dy \, \{J_t^A(x), J_x^B(y)\}_{\rm P} \\ &= \delta^{AB} \lim_{X_1, X_2 \to \infty} \lim_{Y_1, Y_2 \to \infty} [\theta(Y_2 - X_2) - \theta(Y_1 - X_1)] \\ &\quad + (\text{no ambiguity}). \end{aligned} \tag{1.152}$$

Thus the value of the expression in (1.152) depends on the order of limits. A possible resolution is to follow a prescription proposed by MacKay (See the erratum in [17]),

$$X_1 = X_2 \equiv X, \qquad Y_1 = Y_2 \equiv Y. \tag{1.153}$$

However, another ambiguity occurs in computing the Poisson bracket $\{Q^{(1)}, Q^{(1)}\}_{\rm P}$. Typically, it appears from the term

$$\lim_{Y,Y'\to\infty} \int_{-Y}^{Y} dx \int_{-Y}^{Y} dy\,\theta(x-y) \int_{-Y'}^{Y'} dx' \{J_t^B(x)J_t^C(y),\, J_x^A(x')\}_P . \quad (1.154)$$

This term can be evaluated as follows:

$$(1.154) = \lim_{Y,Y'\to\infty} \int_{-Y}^{Y} dx \int_{-Y}^{Y} dy\,\theta(x-y) \int_{-Y'}^{Y'} dx'$$
$$\times \left[\partial_x \delta(x-x')\delta^{AB} \cdot J_t^C(y) + \partial_y \delta(y-x')\delta^{CA} \cdot J_t^B(x) \right]$$
$$+(\text{no ambiguity})$$

$$= \lim_{Y,Y'\to\infty} \left[\delta^{AB} \int_{-Y}^{Y} dy \int_{-Y'}^{Y'} dx' \right.$$
$$\times \left(\theta(Y-y)\delta(Y-x') - \theta(-Y-y)\delta(x'+Y) \right) J_t^C(y)$$
$$+\delta^{CA} \int_{-Y}^{Y} dx \int_{-Y'}^{Y'} dx'$$
$$\left. \times \left(\theta(x-Y)\delta(Y-x') - \theta(x+Y)\delta(Y+x') \right) J_t^B(x) \right]$$
$$+(\text{no ambiguity})$$

$$= \lim_{Y,Y'\to\infty} \left[\delta^{AB} \int_{-Y}^{Y} dy \left(\theta(Y-y) - \theta(-Y-y) \right)\theta(Y'-Y)\,J_t^C(y) \right.$$
$$\left. +\delta^{CA} \int_{-Y}^{Y} dx \left(\theta(x-Y) - \theta(x+Y) \right)\theta(Y'-Y)\,J_t^B(x) \right]$$
$$+(\text{no ambiguity})$$

$$= \lim_{Y,Y'\to\infty} \theta(Y'-Y) \left[\delta^{AB} \int_{-Y}^{Y} dy\,J_t^C(y) - \delta^{CA} \int_{-Y}^{Y} dx\,J_t^B(x) \right] \quad (1.155)$$
$$+(\text{no ambiguity}).$$

Thus the final expression (1.155) depends on the order of limits $Y \to \infty$ and $Y' \to \infty$. If the limit $Y \to \infty$ is taken first, then this term vanishes. But if the limit $Y' \to \infty$ is first, this term becomes

$$\delta^{AB} Q^{(0)C} - \delta^{CA} Q^{(0)B} ,$$

and contributes to the algebra of the charges.

A possible prescription is to choose an appropriate order of the limits so that the Serre relation in the Yangian algebra should be satisfied [17].

Problems

1.1   Lax pair
(a) Show that the Lax pair (1.7) satisfies Eq. (1.4).
(b) Show that the Lax pair (1.9) satisfies Eq. (1.4) as well.

1.2   Flatness condition
Show the flatness conditions (1.24) and (1.25).

1.3   Left-invariant one-form
Compute the expressions in (1.39).

1.4   AdS$_3$ metric
Compute the global AdS$_3$ metric (1.44).

1.5   On-shell flatness of Lax pair
Show the equivalence of (1.89) and (1.90).

1.6   Equation of morion in symmetric coset sigma model
(a) Derive the equation of motion (1.116).
(b) Rewrite (1.116) into (1.117).

1.7   S^2 metric
Compute the S^2 metric (1.136).

1.8   Flatness condition of another Lax pair
Show that (1.145) is satisfied.

1.9   Gauge invariance of Lax pair
Show that (1.145) is invariant under the transformation (1.148).

References

1. E. Abdalla, M.B. Abdalla, D. Rothe, *Non-Perturbative Methods in Two-Dimensional Quantum Field Theory* (World Scientific Publishing Company, Singapore, 1991). [ISBN:978-9810204624]
2. V.I. Arnold, *Mathematical Methods of Classical Mechanics* (Springer, Berlin, 1989). [ISBN:978-1-4757-2063-1]
3. O. Babelon, D. Bernard, M. Talon, *Introduction to Classical Integrable Systems* (Cambridge University Press, Cambridge, 2007). [ISBN:978-0521036702]
4. D. Bernard, An introduction to Yangian symmetries. Int. J. Mod. Phys. B **7**, 3517 (1993). [hep-th/9211133]
5. E. Brezin, C. Itzykson, J. Zinn-Justin, J.B. Zuber, Remarks about the existence of nonlocal charges in two-dimensional models. Phys. Lett. B **82**, 442–444 (1979)
6. A. Das, *Integrable Models* (World Scientific Publishing Company, Singapore, 1989). [ISBN:978-9971509101]

7. F. Delduc, M. Magro, B. Vicedo, Alleviating the non-ultralocality of coset sigma models through a generalized Faddeev-Reshetikhin procedure. JHEP **1208**, 019 (2012). arXiv:1204.0766 [hep-th]

8. V.G. Drinfel'd, Hopf algebras and the quantum Yang-Baxter equation. Sov. Math. Dokl. **32**, 254 (1985)

9. V.G. Drinfel'd, A new realization of Yangians and quantized affine algebras. Sov. Math. Dokl. **36**, 212 (1988)

10. J.M. Evans, M. Hassan, N.J. MacKay, A.J. Mountain, Local conserved charges in principal chiral models. Nucl. Phys. B **561**, 385 (1999). [hep-th/9902008]

11. J.M. Evans, M. Hassan, N.J. MacKay, A.J. Mountain, Conserved charges and supersymmetry in principal chiral and WZW models. Nucl. Phys. B **580**, 605 (2000). [hep-th/0001222]

12. L. Faddeev, N. Reshetikhin, Integrability of the principal chiral field model in $(1 + 1)$-dimension. Ann. Phys. **167**, 227 (1986)

13. L. Faddeev, L. Takhtajan, *Hamiltonian Methods in the Theory of Solitons* (Springer, Berlin, 2007). [ISBN:978-3540698432]

14. M. Luscher, Quantum nonlocal charges and absence of particle production in the two-dimensional nonlinear sigma model. Nucl. Phys. B **135**, 1–19 (1978)

15. M. Luscher, K. Pohlmeyer, Scattering of massless lumps and nonlocal charges in the two-dimensional classical nonlinear sigma model. Nucl. Phys. B **137**, 46–54 (1978)

16. N.J. MacKay, Introduction to Yangian symmetry in integrable field theory. Int. J. Mod. Phys. A **20**, 7189 (2005). [hep-th/0409183]

17. N.J. MacKay, On the classical origins of Yangian symmetry in integrable field theory. Phys. Lett. **B281**, 90–97 (1992). [Erratum: Phys. Lett. **B308** (1993) 444–444]

18. J.M. Maillet, New integrable canonical structures in two-dimensional models. Nucl. Phys. B **269**, 54 (1986)

Chapter 2
Yang–Baxter Sigma Models

Abstract An intriguing subject is to consider integrable deformations of 2D non-linear sigma models such as principal chiral models and symmetric coset sigma models. There had been no systematic way to perform integrable deformations for a long time. However, recently an elegant and systematic way, the so-called Yang–Baxter deformation, was invented by Klimcik. We firstly introduce Yang–Baxter deformations of the principal chiral model and present the associated classical Lax pair. Then the deformation procedure is generalized to the symmetric coset sigma model.

2.1 Classical Yang–Baxter Equation and Linear R-Operator

In order to introduce the Yang–Baxter deformation, it is necessary to make some preparation. A key ingredient in the context of the Yang–Baxter deformation is a linear R-operator from a semi-simple Lie algebra $\mathfrak{g}$ to $\mathfrak{g}$, satisfying the classical Yang–Baxter equation. In this section, we introduce the classical Yang–Baxter equation and then explain some properties of the linear R-operator.

Classical Yang–Baxter Equation

Our starting point is the modified classical Yang–Baxter equation (mCYBE),

$$[R(X), R(Y)] - R([R(X), Y] + [X, R(Y)]) = \omega \cdot [X, Y]. \tag{2.1}$$

This is the equation constraining a linear operator $R: \mathfrak{g} \rightarrow \mathfrak{g}$, where $\mathfrak{g}$ is the semi-simple Lie algebra associated with a semi-simple Lie group G. The constant parameter ω measures the modification and takes one value from ± 1 and 0. Depending on the value of ω, there are three classes of the CYBE like

© The Author(s), under exclusive license to Springer Nature Singapore Pte Ltd. 2021   35
K. Yoshida, *Yang–Baxter Deformation of 2D Non-Linear Sigma Models*,
SpringerBriefs in Mathematical Physics,
https://doi.org/10.1007/978-981-16-1703-4_2

$$\omega = \begin{cases} +1 & \text{(mCYBE of non-split type)} \\ -1 & \text{(mCYBE of split type)} \\ 0 & \text{(homogeneous CYBE)} \end{cases} \tag{2.2}$$

In the original work by Klimcik [7, 8], the mCYBE of non-split type is employed.

When $\omega = 0$, the homogeneous CYBE (hCYBE) is given by

$$[R(X), R(Y)] - R([R(X), Y] + [X, R(Y)]) = 0. \tag{2.3}$$

This is just a special case of the mCYBE, but there is a significant difference between them. In the case of the hCYBE, the right-hand side vanishes, so $R = 0$ is a solution to the hCYBE. But one can immediately understand that $R = 0$ cannot satisfy the mCYBE due to the presence of the modification term.

This difference is crucial and significant when we consider Yang–Baxter deformations of the $AdS_5 \times S^5$ background in the context of string theory. In the case of the mCYBE, both AdS_5 and S^5 must be modified simultaneously and inevitably. However, in the case of the hCYBE, one may consider partial deformations of the geometry, namely, only S^5 or only AdS_5 may be deformed.

It is useful to know the mCYBE in the tensorial notation. For some readers, this expression might be more familiar. The mCYBE in the tensorial notation is given by

$$[r_{12}, r_{13}] + [r_{13}, r_{23}] + [r_{12}, r_{23}] = \omega \cdot [c_{12}, c_{13}]. \tag{2.4}$$

The classical r-matrix,

$$r = \sum_i a_i \otimes b_i \qquad (a_i, b_i \in \mathfrak{g})$$

is defined on a tensor product $\mathfrak{g} \otimes \mathfrak{g}$. The modification term is described with the quadratic Casimir c defined on $\mathfrak{g} \otimes \mathfrak{g}$ as well. The equation (2.4) itself is defined on $\mathfrak{g} \otimes \mathfrak{g} \otimes \mathfrak{g}$. The indices on the classical r-matrix indicate the location of a_i and b_i in $\mathfrak{g} \otimes \mathfrak{g} \otimes \mathfrak{g}$ like

$$r_{12} = \sum_i a_i \otimes b_i \otimes 1, \quad r_{13} = \sum_i a_i \otimes 1 \otimes b_i, \quad r_{23} = \sum_i 1 \otimes a_i \otimes b_i. \tag{2.5}$$

In the following, we are interested in a skew-symmetric R-operator satisfying

$$\langle R(X), Y \rangle = -\langle X, R(Y) \rangle \qquad {}^\forall X, Y \in \mathfrak{g}, \tag{2.6}$$

where the symbol $\langle \ , \ \rangle$ means the non-degenerate inner product for the generators of the semi-simple Lie algebra $\mathfrak{g}$. When a matrix representation is taken, this is basically the trace operation like

$$\langle A, B \rangle = \text{Tr}(AB) . \tag{2.7}$$

Here discussion, we will not take a certain representation for a while and consider in a more general way with this abstract form.

In the tensorial notation, the skew-symmetricity is characterized by

$$r_{ij} = -r_{ji} . \tag{2.8}$$

This property can be realized by taking the following expression:

$$r = \sum_i (a_i \otimes b_i - b_i \otimes a_i) \equiv \sum_i a_i \wedge b_i . \tag{2.9}$$

Relation Between Linear R-Operator and Classical r-Matrix

Under the assumption of the skew-symmetricity, the linear R-operator can be defined by taking the operation on the second site of the tensor product as follows:

$$R(X) \equiv \langle r, 1 \otimes X \rangle = \sum_i \left(a_i \langle b_i, X \rangle - b_i \langle a_i, X \rangle \right), \qquad {}^\forall X \in \mathfrak{g} . \tag{2.10}$$

It is easy to check that the expression (2.10) satisfies the skew-symmetric property (2.6) as follows:

$$\begin{aligned}
\langle R(X), Y \rangle &= \langle \sum_i \left(a_i \langle b_i, X \rangle - b_i \langle a_i, X \rangle \right), Y \rangle \\
&= \sum_i \langle a_i, Y \rangle \langle b_i, X \rangle - \sum_i \langle b_i, Y \rangle \langle a_i, X \rangle \\
&= \langle X, \sum_i b_i \langle a_i, Y \rangle \rangle - \langle X, \sum_i a_i \langle b_i, Y \rangle \rangle \\
&= -\langle X, R(Y) \rangle .
\end{aligned} \tag{2.11}$$

It is also possible to rewrite the mCYBE (2.1) into the tensorial form (2.4). For the detailed computation, see Appendix 1. So far, we have considered the classical version of the Yang–Baxter equation. This can be realized as a semi-classical limit of the quantum Yang–Baxter equation. For this limit, see Appendix 2.

Example: $G = SU(2)$ Case

As a simple example, let us consider the case with $G = SU(2)$ and a classical r-matrix satisfying the mCYBE. In this case, the Lie group is so simple that there is the only one non-trivial solution. This is the classical r-matrix of Drinfeld–Jimbo (DJ) type given by

$$r_{\text{DJ}} = -i(T^+ \otimes T^- - T^- \otimes T^+), \tag{2.12}$$

where $T^\pm$ and T^3 are the generators of $\mathfrak{su}(2)$ and we have set $\omega = 1$. We follow the convention in (1.101) and (1.103) and $T^\pm$ are defined as

$$T^\pm = i(T^1 \pm iT^2), \quad [T^+, T^-] = 2iT^3, \quad [T^\pm, T^3] = \pm T^\pm,$$
$$\text{Tr}(T^+ T^-) = 1, \quad \text{Tr}(T^\pm T^3) = 0. \tag{2.13}$$

Then the associated linear R-operator can be determined by following the definition (2.10). The action of the R-operator is fixed as

$$R(T^+) = -iT^+, \quad R(T^-) = +iT^-, \quad R(T^3) = 0. \tag{2.14}$$

Note here that the above rule indicates that the factor $-i$ should be multiplied for the positive roots, and $+i$ for the negative roots. This rule is the same for the DJ-type classical r-matrix with a higher-rank Lie group as well.

It would be instructive to check that the R-operator (2.14) satisfies the mCYBE. For $X = T^+$ and $Y = T^-$, one can see that

$$[R(T^+), R(T^-)] = [T^+, T^-], \quad R([R(T^+), T^-] + [T^+ R(T^-)]) = 0. \tag{2.15}$$

Thus the linear R-operator (2.14) satisfies the mCYBE with $\omega = 1$. For $X = T^+$ and $Y = T^3$, it is shown that

$$[R(T^+), R(T^3)] = 0, \quad R([R(T^+), T^3] + [T^+, R(T^3)]) = R(-i[T^+, T^3]) = -T^+,$$
$$[T^+, T^3] = T^+. \tag{2.16}$$

Thus in this case, the mCYBE is satisfied as well. Similarly, the mCYBE can be easily checked for the other generators.

2.2 Yang–Baxter Deformation of 2D Principal Chiral Model

In this section, we shall define the Yang–Baxter deformation. The Yang–Baxter deformation was originally proposed by Klimcik [7, 8] as a systematic way to perform integrable deformations of the G-principal chiral model in two dimensions. In the following, let us see Yang–Baxter deformations of the G-principal chiral model action.

2.2.1 Deformed Classical Action

The classical action of the principal chiral model is given in (1.10). Given a linear R-operator with a classical r-matrix, the Yang–Baxter deformation of (1.10) is represented by

$$S^{(\eta)} = -\frac{1}{2}\left(\eta^{\mu\nu} - \epsilon^{\mu\nu}\right)\int d^2x\,\mathrm{Tr}\left(J_\mu \frac{1}{1-\eta R}J_\nu\right). \qquad (2.17)$$

Here η is a real constant parameter[1] and measures the deformation. When $\eta = 0$, the original action (1.10) is reproduced.[2]

Note here that the anti-symmetric tensor $\epsilon^{\mu\nu}$ is included in (2.17) for the consistency with the anti-symmetricity of R. If $\epsilon^{\mu\nu}$ is not contained, the action can be rewritten as

$$-\frac{1}{2}\eta^{\mu\nu}\int d^2x\,\mathrm{Tr}\left(J_\mu \frac{1}{1-\eta^2 R^2}J_\nu\right). \qquad (2.18)$$

This expression is essentially equivalent to the classical action with another *symmetric* operator and thus the anti-symmetricity of R is spoiled down. Hence the ϵ-term is necessary so as to avoid this consequence.

It is convenient to use the light-cone coordinates for base space defined previously. Then the classical action can be rewritten as

$$S^{(\eta)} = \frac{1}{2}\int d^2x\,\mathrm{Tr}\left(J_- \frac{1}{1-\eta R}J_+\right). \qquad (2.19)$$

It is also helpful to introduce deformed currents $A_\pm$ defined as

$$A_\pm \equiv \frac{1}{1 \mp \eta R}J_\pm. \qquad (2.20)$$

Then the classical action is given by

$$S^{(\eta)} = \frac{1}{2}\int d^2x\,\mathrm{Tr}\left(J_- A_+\right) = \frac{1}{2}\int d^2x\,\mathrm{Tr}\left(A_- J_+\right). \qquad (2.21)$$

---

[1] The scalar η is the deformation parameter and the symmetric tensor $\eta_{\mu\nu}$ is the metric of 2D Minkowski spacetime. Sorry for this misleading notation but this is the standard one in this subject.
[2] The contribution with $\epsilon^{\mu\nu}$ vanishes because $\epsilon^{\mu\nu}\mathrm{Tr}(J_\mu J_\nu) = 0$.

Symmetry

The symmetry of the original principal chiral model is given by a product of the left global G-symmetry G_L and the right global G-symmetry G_R, like $G_L \times G_R$. After performing the Yang–Baxter deformation as in (2.17), the right global symmetry G_R is broken. This is just because the linear R-operator acts on the left-invariant current, which is associated with the global right symmetry, by construction of the deformed action. It is possible to break the global left symmetry G_L by combining the adjoint operations as described in detail later. This procedure will play an important role in the symmetric case.

Equation of Motion

To obtain the equation of motion, let us take a variation of $g \in G$ as $\delta g = g\epsilon$, where $\epsilon \in G$ as well. The associated variation of the left-invariant current J_μ is given by

$$\delta J_\mu = \partial_\mu \epsilon + [J_\mu, \epsilon] . \tag{2.22}$$

Then the variation of the classical action is evaluated as

$$\delta S^{(\eta)} = -\frac{1}{2} \int d^2x \, \mathrm{Tr}(\mathcal{E} \epsilon) \tag{2.23}$$

where $\mathcal{E}$ is defined as

$$\mathcal{E} \equiv \partial_+ A_- + \partial_- A_+ - \eta \left([R(A_+), A_-] + [A_+, R(A_-)] \right) . \tag{2.24}$$

Thus the equation of motion turns out to be $\mathcal{E} = 0$. For the detail of the derivation of (2.24), see Appendix 3.

Flatness Condition

Let us introduce the following quantity:

$$\mathcal{Z} \equiv \partial_+ J_- - \partial_- J_+ + [J_+, J_-] . \tag{2.25}$$

By the definition of J_μ, this quantity should vanish i.e., $\mathcal{Z} = 0$. From (2.20),

$$J_\pm = (1 \mp \eta R) A_\pm . \tag{2.26}$$

By putting (2.26) into (2.25), we obtain the following expression:

$$\mathcal{Z} = \partial_+ A_- - \partial_- A_+ - \eta\left([R(A_+), A_-] - [A_+, R(A_-)]\right)$$
$$+[A_+, A_-] - \eta^2 \text{YBE}(A_+, A_-) + \eta R(\mathcal{E}), \tag{2.27}$$

where $\text{YBE}(X, Y)$ is defined as

$$\text{YBE}(X, Y) \equiv [R(X), R(Y)] - R([R(X), Y] + [X, R(Y)]). \tag{2.28}$$

By using the mCYBE (2.1), the quantity $\mathcal{Z}$ is rewritten as

$$\mathcal{Z} = \partial_+ A_- - \partial_- A_+ - \eta\left([R(A_+), A_-] - [A_+, R(A_-)]\right)$$
$$+(1 - \eta^2\omega)[A_+, A_-] + \eta R(\mathcal{E}). \tag{2.29}$$

Lax Pair

The deformed Lax pair is given by

$$\mathcal{L}_\pm^{(\eta,\omega)}(x; \lambda) = \left(\frac{1 \mp \eta^2\omega\lambda}{1 \pm \lambda} \mp \eta R\right) A_\pm$$
$$= \frac{1}{1 \pm \lambda}\left(1 \mp \frac{\eta\lambda(\eta\omega \pm R)}{1 \pm \eta R}\right) J_\pm \tag{2.30}$$

with a spectral parameter $\lambda \in \mathbb{C}$. When $\eta = 0$, the undeformed Lax pair (1.85) is reproduced.

Indeed, the equation of motion $\mathcal{E} = 0$ and the flatness condition $\mathcal{Z} = 0$ are equivalent to the flatness condition of the Lax pair $\mathcal{L}_\mu^{(\eta,\omega)}$,

$$\partial_+ \mathcal{L}_-^{(\eta,\omega)} - \partial_- \mathcal{L}_+^{(\eta,\omega)} + [\mathcal{L}_+^{(\eta,\omega)}, \mathcal{L}_-^{(\eta,\omega)}] = 0. \tag{2.31}$$

Thus, the Yang–Baxter deformed principal chiral model is also classically integrable in the sense of kinematical integrability.

Proof of the Flatness Condition of Lax Pair

It is a bit messy to check the flatness condition of $\mathcal{L}^{(\eta,\omega)}$. Here, let us start from a bit general expression of the Lax pair with two arbitrary functions,

$$\mathcal{L}_\pm^{(\eta,\omega)}(x; \lambda) = \left(\frac{F \pm G\lambda}{1 \pm \lambda} \mp \eta R\right) A_\pm. \tag{2.32}$$

Here F and G are unknown functions of η and ω to be determined by imposing the flatness condition on $\mathcal{L}^{(\eta,\omega)}$. Suppose that these functions do not depend on the spectral parameter λ and the world-sheet coordinates.

Then one can evaluate the following quantity:

$$
\begin{aligned}
\partial_+ & \mathcal{L}_-^{(\eta,\omega)} - \partial_- \mathcal{L}_+^{(\eta,\omega)} + [\mathcal{L}_+^{(\eta,\omega)}, \mathcal{L}_-^{(\eta,\omega)}] \\
= & \frac{1}{1-\lambda^2} \left[FZ + (F-1)\left((F+\eta^2\omega)[A_+, A_-] - \eta R(\mathcal{E}) \right) \right] \\
& + \frac{\lambda}{1-\lambda^2}(F-G)\mathcal{E} \\
& - \frac{\lambda^2}{1-\lambda^2} \left[GZ + (G-1)\left((G+\eta^2\omega)[A_+, A_-] - \eta R(\mathcal{E}) \right) \right] .
\end{aligned}
\tag{2.33}
$$

Thus, to see the flatness condition, there are four possibilities for the pair F and G:

$$
(F, G) = (1, 1), \quad (1, -\eta^2\omega), \quad (-\eta^2\omega, 1), \quad (-\eta^2\omega, -\eta^2\omega) .
\tag{2.34}
$$

For the first case, the Lax pair is reduced to $J_\pm$ and hence this corresponds to a trivial solution. In the fourth case, $\mathcal{E}$ itself vanishes from (2.33), though $R(\mathcal{E})$ remains. Hence the information on $\mathcal{E}$ cannot be reproduced, so this choice should be discarded. In summary, the relevant choices are

$$
(F, G) = (1, -\eta^2\omega), \quad (-\eta^2\omega, 1) .
\tag{2.35}
$$

The first choice is nothing but (2.30). The second choice is equivalent to the first one by replacing λ by $1/\lambda$.

Example 1: $G = SU(2)$ Case with mCYBE

It would be a good exercise to examine the Yang–Baxter deformed action (2.17) or equivalently (2.21) for a simple case with $G = SU(2)$. In this case the non-trivial classical r-matrix satisfying the mCYBE is uniquely determined and it is the classical r-matrix of Drinfeld–Jimbo type in (2.12). Then the transformation rule of the linear R-operator is given in (2.14).

Both J_μ and A_μ take values in $\mathfrak{su}(2)$ and can be expanded as

$$
J_\mu = J_\mu^+ T^- + J_\mu^- T^+ + J_\mu^3 T^3 ,
\tag{2.36}
$$

$$
A_\mu = A_\mu^+ T^- + A_\mu^- T^+ + A_\mu^3 T^3 .
\tag{2.37}
$$

It is also useful to see the relations:

$$
J_\mu^\pm = -\frac{i}{2}(J_\mu^1 \pm iJ_\mu^2), \quad J_\mu^+ J_\nu^- = -\frac{1}{4}\left[J_\mu^1 J_\nu^1 + J_\mu^2 J_\nu^2 - i(J_\mu^1 J_\nu^2 - J_\nu^1 J_\mu^2) \right] .
\tag{2.38}
$$

Recall that the action (2.17) is now written in terms of J and A. By multiplying the factor $1 \mp \eta R$ to $A_{\pm}$, $J_{\pm}$ can be expressed in terms of $A_{\pm}^{\pm}$ and $A_{\pm}^{3}$ as

$$J_{\pm} = (1 \mp \eta R)A_{\pm}$$
$$= (1 \mp i\eta)A_{\pm}^{+}T^{-} + (1 \pm i\eta)A_{\pm}^{-}T^{+} + A_{\pm}^{3}T^{3}. \qquad (2.39)$$

Here in the last line, we have used the transformation rule of the linear R-operator (2.14). Then, by comparing (2.39) with (2.36), one finds that $A_{\pm}^{\pm}$ and $A_{\pm}^{3}$ can be expressed as

$$A_{\pm}^{+} = \frac{J_{\pm}^{+}}{1 \mp i\eta}, \qquad A_{\pm}^{-} = \frac{J_{\pm}^{-}}{1 \pm i\eta}, \qquad A_{\pm}^{3} = J_{\pm}^{3}. \qquad (2.40)$$

Thus, the projected current is given by

$$A_{\pm} = \frac{1}{1 \mp i\eta}J_{\pm}^{+}T^{-} + \frac{1}{1 \pm i\eta}J_{\pm}^{-}T^{+} + J_{\pm}^{3}T^{3}. \qquad (2.41)$$

By using the expression (2.41), the classical action (2.21) can be rewritten as

$$S^{(\eta)} = \frac{1}{2}\int d^{2}x \, \mathrm{Tr}(J_{-}A_{+})$$
$$= \frac{1}{2}\int d^{2}x \left[\frac{1}{1+i\eta}J_{-}^{+}J_{+}^{-} + \frac{1}{1-i\eta}J_{-}^{-}J_{+}^{+} - \frac{1}{2}J_{+}^{3}J_{-}^{3} \right]$$
$$= \frac{1}{2}\int d^{2}x \left[(J_{t}^{+}J_{t}^{-} - J_{x}^{+}J_{x}^{-})\left(\frac{1}{1+i\eta} + \frac{1}{1-i\eta} \right) \right.$$
$$\left. + (J_{t}^{+}J_{x}^{-} - J_{x}^{+}J_{t}^{-})\left(\frac{1}{1+i\eta} - \frac{1}{1-i\eta} \right) - \frac{1}{2}(J_{t}^{3}J_{t}^{3} - J_{x}^{3}J_{x}^{3}) \right]$$
$$= -\frac{1}{1+\eta^{2}}\eta^{\mu\nu}\int d^{2}x \left[J_{\mu}^{+}J_{\nu}^{-} - \frac{1}{4}(1+\eta^{2})J_{\mu}^{3}J_{\nu}^{3} \right]$$
$$- \frac{2i\eta}{1+\eta^{2}}\epsilon^{\mu\nu}\int d^{2}x \, J_{\mu}^{+}J_{\nu}^{-}. \qquad (2.42)$$

Finally, by using the relation in (2.38), the classical action is given by

$$S^{(\eta)} = -\frac{1}{2}\eta^{\mu\nu}\left(\frac{1}{1+\eta^{2}} \right)\int d^{2}x \left[\mathrm{Tr}(J_{\mu}J_{\nu}) - 2\eta^{2}\mathrm{Tr}(T^{3}J_{\mu})\mathrm{Tr}(T^{3}J_{\nu}) \right]$$
$$+ \frac{1}{2}\epsilon^{\mu\nu}\left(\frac{1}{1+\eta^{2}} \right)\int d^{2}x \, \eta(J_{\mu}^{1}J_{\nu}^{2} - J_{\nu}^{1}J_{\mu}^{2}). \qquad (2.43)$$

Note here that the anti-symmetric part appears in addition to the symmetric part. The anti-symmetric part gives rise to an anti-symmetric two-form B_{2} which is turned on in the target space, while the symmetric part provides the target-space metric.

From the symmetric part, by following the normalization so as to reproduce the standard metric of S^3, the metric is given by

$$ds^2 = -\frac{1}{2}\mathrm{Tr}(JJ) + \eta^2\mathrm{Tr}(T^3 J)\mathrm{Tr}(T^3 J)$$
$$= \frac{1}{4}\left[d\theta^2 + \sin^2\theta d\phi^2 + (1+\eta^2)(d\psi + \cos\theta d\phi)^2\right], \qquad (2.44)$$

where we have used the expressions in (1.39). This is the metric of squashed S^3.

To read off the anti-symmetric two-form B_2, we need to fix the normalization of the coupling to B_2. Including the sigma-model part, the standard action is given by

$$S_{\sigma M} = -\frac{1}{2}\int d^2x\, \eta^{\mu\nu}G_{MN}\partial_\mu X^M \partial_\nu X^N$$
$$+\frac{1}{2}\int d^2x\, \epsilon^{\mu\nu} B_{MN}\partial_\mu X^M \partial_\nu X^N. \qquad (2.45)$$

Taking account of the normalization of the sigma-model part, the coupling to B_{MN} is evaluated as follows:

$$B_{MN}\partial_\mu X^M \partial_\nu X^N = -\frac{\eta}{2}(J_\mu^1 J_\nu^2 - J_\nu^1 J_\mu^2)$$
$$= -\frac{\eta}{2}\sin\theta(\partial_\mu\theta\partial_\nu\phi - \partial_\nu\theta\partial_\mu\phi). \qquad (2.46)$$

Thus, the anti-symmetric two-form B_2 is given by

$$B_2 = \frac{\eta}{2}d\phi \wedge \sin\theta d\theta. \qquad (2.47)$$

Example 2: $G = SL(2)$ Case with hCYBE

Next, let us consider the following skew-symmetric classical r-matrix:

$$r = T^2 \otimes T^- - T^- \otimes T^2, \qquad (2.48)$$

where we follow the convention for the generators in (1.41) and (1.46). The associated linear R-operator is obtained as

$$R(T^+) = -T^2, \quad R(T^-) = 0, \quad R(T^2) = -\frac{1}{2}T^-. \qquad (2.49)$$

Note here that this R-operator exhibits the nilpotent property:

$$R^3 = 0. \qquad (2.50)$$

The left-invariant one-form J is expanded as

$$J = J^+ T^- + J^- T^+ + J^2\, T^2\,. \tag{2.51}$$

Then, due to the property (2.50), the projected current $A_\pm$ can be expanded as

$$A_\pm = (1 \pm \eta R + \eta^2 R^2) J_\pm\,. \tag{2.52}$$

Since

$$R(J_\pm) = -J_\pm^- T^2 - \frac{1}{2} J_\pm^2\, T^-\,, \quad R^2(J_\pm) = \frac{1}{2} J_\pm^- T^-\,, \tag{2.53}$$

the projected current $A_\pm$ can be expressed as

$$A_\pm = \left(J_\pm^+ \mp \frac{1}{2}\eta J_\pm^2 + \frac{1}{2}\eta^2 J_\pm^- \right) T^- + J_\pm^- T^+ + \left(J_\pm^2 \mp \eta J_\pm^- \right) T^2\,. \tag{2.54}$$

By using this expression, the deformed action can be evaluated as follows:

$$
\begin{aligned}
S^{(\eta)} &= \frac{1}{2} \int d^2x\, \mathrm{Tr}(J_- A_+) \\
&= \int d^2x \left[-\frac{1}{2}(J_-^- J_+^+ + J_-^+ J_+^-) + \frac{1}{4} J_-^2 J_+^2 \right. \\
&\qquad\qquad \left. + \frac{1}{4}\eta \left(J_-^- J_+^2 - J_-^2 J_+^- \right) - \frac{1}{4}\eta^2 J_-^- J_+^- \right] \\
&= -\frac{1}{2}\eta^{\mu\nu} \int d^2x \left(\mathrm{Tr}(J_\mu J_\nu) - \frac{1}{2}\eta^2 \mathrm{Tr}(T^+ J_\mu)\mathrm{Tr}(T^+ J_\nu) \right) \\
&\qquad - \frac{1}{2}\epsilon^{\mu\nu} \int d^2x\, \eta\, J_\mu^2 J_\nu^-\,.
\end{aligned}
\tag{2.55}
$$

By using the normalization in the case of Poincaré AdS$_3$ in (1.51), the target-space metric is given by

$$
\begin{aligned}
ds^2 &= \frac{1}{2}\mathrm{Tr}(JJ) - \frac{1}{4}\eta^2 \mathrm{Tr}(T^+ J)\mathrm{Tr}(T^+ J) \\
&= \frac{-4dx^+ dx^- + dz^2}{z^2} - \eta^2 \frac{(dx^+)^2}{z^4}\,.
\end{aligned}
\tag{2.56}
$$

This is 3D Schrödinger spacetime.

The anti-symmetric two-form B_{MN} is given by

$$B_{MN}\partial_\mu X^M \partial_\nu X^N = -\frac{\eta}{4}(J_\mu^2 J_\nu^- - J_\nu^2 J_\mu^-)$$

$$= -\frac{\eta}{2}\left(\partial_\mu x^+ \partial_\nu\left(\frac{1}{z^2}\right) - \partial_\nu x^+ \partial_\mu\left(\frac{1}{z^2}\right)\right). \qquad (2.57)$$

Thus we obtain that

$$B_2 = -\frac{\eta}{2}dx^+ \wedge d\left(\frac{1}{z^2}\right). \qquad (2.58)$$

2.3   Yang–Baxter Deformation of Symmetric Coset Model

In this section, let us consider the Yang–Baxter deformation of the symmetric coset sigma model. The symmetric coset is represented by $M = G/H$ again.

2.3.1   Deformed Classical Action

The classical action has been introduced in (1.105). The Yang–Baxter deformation of (1.105) is represented by the following Lagrangian:

$$S_{\text{symm}}^{(\eta)} = -\frac{1}{2}(\eta^{\mu\nu} - \epsilon^{\mu\nu})\int d^2 x \, \text{Tr}\left(J_\mu P \circ \frac{1}{1 - \eta R_g \circ P}J_\nu\right). \qquad (2.59)$$

Here P is the projection introduced in (1.104) and R_g is a dressed R-operator defined as

$$R_g(X) \equiv g^{-1}R(gXg^{-1})g \qquad \text{with} \quad g \in G \text{ and } X \in \mathfrak{g}. \qquad (2.60)$$

The linear R-operator should satisfy the mCYBE (2.1) or the hCYBE (2.3), again.

The dressed operator R_g means a chain operation composed of the adjoint operation with g, the action of R and then the inverse adjoint operation like

$$R_g = \text{Ad}_{g^{-1}} \circ R \circ \text{Ad}_g. \qquad (2.61)$$

What is the meaning of this chain operation? It indicates that the R-operator breaks the global left symmetry.

Recall that the R-operator broke the global right symmetry in the case of the principal chiral model. This is just because the R-operator acts on the left-invariant current, corresponding to the Noether current of the global right symmetry. In the case of the symmetric coset sigma model, one has to devise how to break the global left symmetry, rather than the global right symmetry because the global left symmetry exists while there is no global right symmetry. This trick is nothing but the chain operation (2.60) or equivalently (2.61). By taking the first adjoint operation, the left-invariant current is mapped to the right-invariant current, which is associated with the global left symmetry. Then a linear R-operator acts on the right-invariant current, and then the global left symmetry is broken to a smaller subgroup. Finally, taking the inverse adjoint operation recovers the description based on the left-invariant current. This is the meaning of the dressed R-operator R_g.

As a matter of course, by using the dressed R-operator R_g in the principal chiral model, the global left symmetry can be broken as well.

It is convenient to introduce a projected current,

$$\tilde{A}_\pm \equiv \frac{1}{1 \mp \eta R_g \circ P} J_\pm \, . \tag{2.62}$$

Then the classical action is given by

$$S^{(\eta)} = \frac{1}{2} \int d^2x \, \text{Tr} \left(J_- P(\tilde{A}_+) \right) = \frac{1}{2} \int d^2x \, \text{Tr} \left(P(\tilde{A}_-) J_+ \right) \, . \tag{2.63}$$

Equation of Motion

By taking a variation of the classical action (2.59) or equivalently (2.63), one can show that the equation of motion is given by

$$\tilde{\mathcal{E}} = 0 \, ,$$
$$\tilde{\mathcal{E}} \equiv \partial_+ P(\tilde{A}_-) + \partial_- P(\tilde{A}_+) + \left[\tilde{A}_+, P(\tilde{A}_-) \right] + \left[\tilde{A}_-, P(\tilde{A}_+) \right] \, . \tag{2.64}$$

For the detail of the derivation, see Appendix 4.

Flatness Condition

Let us evaluate the flatness condition

$$\tilde{\mathcal{Z}} \equiv \partial_+ J_- - \partial_- J_+ + [J_+, J_-] = 0 \tag{2.65}$$

in terms of the projected current $\tilde{A}_\pm$. By noticing that

$$J_\pm = (1 \mp \eta R_g \circ P) \tilde{A}_\pm \, , \tag{2.66}$$

the quantity $\tilde{\mathcal{Z}}$ can be rewritten as

$$\tilde{\mathcal{Z}} = \partial_+ \tilde{A}_- - \partial_- \tilde{A}_+ + [\tilde{A}_+, \tilde{A}_-] + \eta R_g(\tilde{\mathcal{E}}) + \eta^2 \text{YBE}_g(P(\tilde{A}_+), P(\tilde{A}_-)). \quad (2.67)$$

Here we have introduced a new symbol,

$$\text{YBE}_g \equiv [R_g(X), R_g(Y)] - R_g([R_g(X), Y] + [X, R_g(Y)]). \quad (2.68)$$

Note that when R is a solution to the mCYBE, R_g also satisfies the mCYBE. Hence when R satisfies the mCYBE, we obtain

$$\text{YBE}_g(X, Y) = \omega \cdot [X, Y], \quad (2.69)$$

and the expression of $\tilde{\mathcal{Z}}$ further reduces to

$$\tilde{\mathcal{Z}} = \partial_+ \tilde{A}_- - \partial_- \tilde{A}_+ + [\tilde{A}_+, \tilde{A}_-] + \eta R_g(\tilde{\mathcal{E}}) + \eta^2 \omega [P(\tilde{A}_+), P(\tilde{A}_-)]. \quad (2.70)$$

2.3.2 Lax Pair

The deformed Lax pair is given by

$$\tilde{\mathcal{L}}_\pm^{(\eta,\omega)}(x, \lambda) \equiv \tilde{A}_\pm^{(0)} + \lambda^{\pm 1} \sqrt{1 + \eta^2 \omega}\, \tilde{A}_\pm^{(1)}, \quad (2.71)$$

where $\tilde{A}^0$ and $\tilde{A}^{(1)}$ are the grade zero and one components of the projected current $\tilde{A}$, respectively. In particular, $\tilde{A}^{(1)} = P(\tilde{A})$ and $\tilde{A}^{(0)} = \tilde{A} - P(\tilde{A})$. $\lambda (\in \mathbb{C})$ is a spectral parameter again. When $\eta = 0$, the undeformed Lax pair (1.144) is reproduced.

Note here that in the case of hCYBE i.e., ($\omega = 0$), the Lax pair is given by replacing the left-invariant one-form J by the projected current $\tilde{A}$.

The flatness condition

$$\partial_+ \tilde{\mathcal{L}}_-^{(\eta,\omega)} - \partial_- \tilde{\mathcal{L}}_+^{(\eta,\omega)} + [\tilde{\mathcal{L}}_+^{(\eta,\omega)}, \tilde{\mathcal{L}}_-^{(\eta,\omega)}] = 0 \quad (2.72)$$

is equivalent to $\tilde{\mathcal{E}} = 0$ and $\tilde{\mathcal{Z}} = 0$.

Proof of the Flatness Condition

Let us show the flatness condition (2.72).

We start from a bit general expression,

$$\tilde{\mathcal{L}}_\pm^{(\eta,\omega)} = \tilde{A}_\pm^{(0)} + \lambda^{\pm 1} \tilde{G} \tilde{A}_\pm^{(1)}, \quad (2.73)$$

where $\tilde{G}$ is an unknown function of η and ω to be determined by the flatness condition. Note that this function depends on neither the spectral parameter λ nor the world-sheet coordinates.

Then we obtain the following expression:

$$
\begin{aligned}
&\partial_+ \tilde{\mathcal{L}}_-^{(\eta,\omega)} - \partial_- \tilde{\mathcal{L}}_+^{(\eta,\omega)} + [\tilde{\mathcal{L}}_+^{(\eta,\omega)}, \tilde{\mathcal{L}}_-^{(\eta,\omega)}] \\
&= \partial_+ \tilde{A}_-^{(0)} - \partial_- \tilde{A}_+^{(0)} + [\tilde{A}_+^{(0)}, \tilde{A}_-^{(0)}] + \tilde{G}^2 [\tilde{A}_+^{(1)}, \tilde{A}_-^{(1)}] \\
&\quad - \lambda \tilde{G} \left(\partial_- \tilde{A}_+^{(1)} - [\tilde{A}_+^{(1)}, \tilde{A}_-^{(0)}] \right) + \lambda^{-1} \tilde{G} \left(\partial_+ \tilde{A}_-^{(1)} + [\tilde{A}_+^{(0)}, \tilde{A}_-^{(1)}] \right). \quad (2.74)
\end{aligned}
$$

By rewriting the equation of motion (2.64), we obtain

$$
\begin{aligned}
0 &= \partial_+ \tilde{A}_-^{(1)} + \partial_- \tilde{A}_+^{(1)} + [\tilde{A}_+^{(0)}, \tilde{A}_-^{(1)}] + [\tilde{A}_+^{(1)}, \tilde{A}_-^{(1)}] \\
&\quad + [\tilde{A}_-^{(0)}, \tilde{A}_+^{(1)}] + [\tilde{A}_-^{(1)}, \tilde{A}_+^{(1)}]. \quad (2.75)
\end{aligned}
$$

By acting the projection operator P to the above equation, the components with the grade zero vanish, and the following equation is obtained:

$$
0 = \partial_+ \tilde{A}_-^{(1)} + \partial_- \tilde{A}_+^{(1)} + [\tilde{A}_+^{(0)}, \tilde{A}_-^{(1)}] - [\tilde{A}_+^{(1)}, \tilde{A}_-^{(0)}]. \quad (2.76)
$$

Then, after using $\tilde{\mathcal{E}} = 0$, based on the grading, the flatness condition $\tilde{\mathcal{Z}} = 0$ can be decomposed into

$$
\partial_+ \tilde{A}_-^{(0)} - \partial_- \tilde{A}_+^{(0)} + [\tilde{A}_+^{(0)}, \tilde{A}_-^{(0)}] + (1 + \eta^2 \omega)[\tilde{A}_+^{(1)}, \tilde{A}_-^{(1)}] = 0, \quad (2.77)
$$
$$
\partial_+ \tilde{A}_-^{(1)} - \partial_- \tilde{A}_+^{(1)} + [\tilde{A}_+^{(0)}, \tilde{A}_-^{(1)}] + [\tilde{A}_+^{(1)}, \tilde{A}_-^{(0)}] = 0. \quad (2.78)
$$

By combining (2.76) with (2.78), the following two equations are obtained:

$$
\partial_+ \tilde{A}_-^{(1)} + [\tilde{A}_+^{(0)}, \tilde{A}_-^{(1)}] = 0, \quad (2.79)
$$
$$
\partial_- \tilde{A}_+^{(1)} - [\tilde{A}_+^{(0)}, \tilde{A}_-^{(0)}] = 0. \quad (2.80)
$$

Finally, using (2.77), (2.79) and (2.80), we obtain

$$
\begin{aligned}
&\partial_+ \tilde{\mathcal{L}}_-^{(\eta,\omega)} - \partial_- \tilde{\mathcal{L}}_+^{(\eta,\omega)} + [\tilde{\mathcal{L}}_+^{(\eta,\omega)}, \tilde{\mathcal{L}}_-^{(\eta,\omega)}] \\
&= \left(\tilde{G}^2 - (1 + \eta^2 \omega) \right) [\tilde{A}_+^{(1)}, \tilde{A}_-^{(1)}]. \quad (2.81)
\end{aligned}
$$

Hence, under the condition

$$
\tilde{G} = \pm \sqrt{1 + \eta^2 \omega}, \quad (2.82)
$$

the flatness condition is satisfied. The signature of G can be absorbed by redefining the spectral parameter λ.

2.4 History of the Development of the Yang–Baxter Deformation

The following is the history of the development of the Yang–Baxter deformation. It is mainly composed of two classes depending on the mCYBE or the hCYBE.

(1) **The class based on the mCYBE**

    (a) Principal chiral model
        C. Klimcik [7, 8]

    (b) Symmetric coset sigma model
        F. Delduc, M. Magro and B. Vicedo [2]

    (c) Type IIB superstring on $AdS_5 \times S^5$
        F. Delduc, M. Magro and B. Vicedo [3]

(2) **The class based on the hCYBE**

    (a) Principal chiral model
        T. Matsumoto and K. Yoshida [10]

    (b) Symmetric coset sigma model
        T. Matsumoto and K. Yoshida [10]

    (c) Type IIB superstring on $AdS_5 \times S^5$
        I. Kawaguchi, T. Matsumoto and K. Yoshida [6]

As a matter of course, the Yang–Baxter deformation was originally invented for the principal chiral model by Klimcik [7, 8]. Then it was generalized to the symmetric coset sigma model by Delduc, Magro and Vicedo [2]. Then immediately they succeeded in applying their result to type IIB string theory on the $AdS_5 \times S^5$ background [3]. It is well known that type IIB string theory on the $AdS_5 \times S^5$ background is classically integrable [1] (for the related works, see [4, 5]).

As the third one, one may consider the bi-Yang–Baxter deformation [8, 9]. This deformation is applicable only for the principal chiral model, and both left and right symmetries, G_L and G_R are independently deformed. The classical integrability of the bi-Yang–Baxter sigma model has been shown in [9] by explicitly constructing the Lax pair.

Appendices

Appendix 1: Rewriting the CYBE

Let us here rewrite the classical Yang–Baxter equation in terms of the associated linear R-operator to the one in the tensorial notation.

Suppose that a classical r-matrix in the tensorial notation is written as

$$r = \sum_i a_i \otimes b_i \quad \in \mathfrak{g} \otimes \mathfrak{g}. \tag{2.83}$$

Here $\mathfrak{g}$ is the Lie algebra associated with a Lie group G. Then the mCYBE in the tensorial notation is given by

$$[r_{12}, r_{13}] + [r_{13}, r_{23}] + [r_{12}, r_{23}] = \omega \cdot [c_{12}, c_{13}], \tag{2.84}$$

where the above equation is defined on the triple tensor product $\mathfrak{g} \otimes \mathfrak{g} \otimes \mathfrak{g}$ and

$$r_{12} = \sum_i a_i \otimes b_i \otimes 1, \quad r_{23} = \sum_i 1 \otimes a_i \otimes b_i, \quad r_{13} = \sum_i a_i \otimes 1 \otimes b_i. \tag{2.85}$$

The subscripts with r indicate the location of a_i and b_i. The left and right subscripts are the site number of a_i and b_i, respectively. Then c_{12} and c_{13} are quadratic Casimir operators defined as

$$c_{12} = \sum_{a=1}^{\dim(\mathfrak{g})} T_a \otimes T_a \otimes 1, \quad c_{13} = \sum_{a=1}^{\dim(\mathfrak{g})} T_a \otimes 1 \otimes T_a, \tag{2.86}$$

where T_a's are the orthonormalized generators of the Lie algebra $\mathfrak{g}$ as $\langle T_a, T_b \rangle = \delta_{ab}$.

We would like to show the following claim. Given a skew-symmetric classical r-matrix,[3] such that $r_{21} = \sum_i b_i \otimes a_i = -\sum_i a_i \otimes b_i = -r_{12}$, the mCYBE in the tensorial notation (2.84) can be derived from

$$[R(X), R(Y)] - R([R(X), Y] + [X, R(Y)]) = \omega \cdot [X, Y] \tag{2.87}$$

in terms of the associated linear R-operator.

First of all, the first term of (2.87) can be evaluated as

---

[3]The skew-symmetricity realized in (2.9) is here described in a different notation for simplicity.

$$\begin{aligned}
[R(X), R(Y)] &= \sum_{i,j} [a_i \langle b_i, X \rangle, a_j \langle b_j, Y \rangle] \\
&= \sum_{i,j} \langle b_i, X \rangle \langle b_j, Y \rangle [a_i, a_j] \\
&= \langle \sum_{i,j} [a_i, a_j] \otimes b_i \otimes b_j, 1 \otimes X \otimes Y \rangle \\
&= \langle [\sum_i a_i \otimes b_i \otimes 1, \sum_j a_j \otimes 1 \otimes b_j], 1 \otimes X \otimes Y \rangle \\
&= \langle [r_{12}, r_{13}], 1 \otimes X \otimes Y \rangle .
\end{aligned}$$
(2.88)

Similarly, the second term of (2.87) is rewritten as

$$\begin{aligned}
- R([R(X), Y]) &= -R \left([\sum_i a_i \langle b_i, X \rangle, Y] \right) \\
&= -R \left(\sum_i [a_i, Y] \langle b_i, X \rangle \right) \\
&= - \sum_{i,j} a_j \langle b_j, [a_i, Y] \rangle \langle b_i, X \rangle \\
&= \sum_{i,j} a_j \langle [a_i, b_j], Y \rangle \langle b_i, X \rangle \\
&= \langle \sum_{i,j} a_j \otimes b_j \otimes [a_i, b_j], 1 \otimes X \otimes Y \rangle \\
&= \langle [\sum_i 1 \otimes b_i \otimes a_i, \sum_j a_j \otimes 1 \otimes b_j], 1 \otimes X \otimes Y \rangle \\
&= \langle [r_{32}, r_{13}], 1 \otimes X \otimes Y \rangle \\
&= \langle [r_{13}, r_{23}], 1 \otimes X \otimes Y \rangle .
\end{aligned}$$
(2.89)

Then, the third term of (2.87) is evaluated as

$$- R([X, R(Y)]) = -R\left([X, \sum_i a_i \langle b_i, Y\rangle]\right)$$

$$= -R \sum_i [X, a_i]\langle b_i, Y\rangle$$

$$= -\sum_{i,j} a_j \langle b_j, [X, a_i]\rangle\langle b_i, Y\rangle$$

$$= -\sum_{i,j} a_j \langle [a_i, b_j], X\rangle\langle b_i, Y\rangle$$

$$= -\langle \sum_{i,j} a_j \otimes [a_i, b_j] \otimes b_i, 1 \otimes X \otimes Y\rangle$$

$$= -\langle [\sum_i 1 \otimes a_i \otimes b_i, \sum_j a_j \otimes b_j \otimes 1], 1 \otimes X \otimes Y\rangle$$

$$= -\langle [r_{23}, r_{12}], 1 \otimes X \otimes Y\rangle$$

$$= \langle [r_{12}, r_{23}], 1 \otimes X \otimes Y\rangle. \tag{2.90}$$

Finally, let us rewrite the modification term:

$$[X, Y] = \sum_{a,b} [\langle X, T_a\rangle T_a, \langle Y, T_b\rangle T_b]$$

$$= \sum_{a,b} [T_a, T_b]\langle X, T_a\rangle\langle Y, T_b\rangle$$

$$= \langle \sum_{a,b} [T_a, T_b] \otimes T_a \otimes T_b, 1 \otimes X \otimes Y\rangle$$

$$= \langle \sum_{a,b} [T_a \otimes T_a \otimes 1, T_b \otimes 1 \otimes T_b], 1 \otimes X \otimes Y\rangle$$

$$= \langle [c_{12}, c_{13}], 1 \otimes X \otimes Y\rangle. \tag{2.91}$$

By summing up (2.88), (2.89), (2.90) and (2.91), it has been shown that

$$0 = [R(X), R(Y)] - R([R(X), Y] + [X, R(Y)]) - \omega \cdot [X, Y]$$

$$= \langle [r_{12}, r_{13}] + [r_{13}, r_{23}] + [r_{12}, r_{23}] - \omega \cdot [c_{12}, c_{13}], 1 \otimes X \otimes Y\rangle. \tag{2.92}$$

The last quantity should vanish for arbitrary $1 \otimes X \otimes Y$ and hence the mCYBE in the tensorial notation (2.84) has been derived.

Appendix 2: Semi-classical Limit of Quantum Yang–Baxter Equation

The quantum Yang–Baxter equation (QYBE) (Fig. 2.1) is defined as

$$R_{12}R_{13}R_{23} = R_{23}R_{13}R_{12}, \tag{2.93}$$

where R_{ij} is a quantum R-matrix defined on a tensor product $\mathfrak{g} \otimes \mathfrak{g}$.

The hCYBE can be realized as a semi-classical limit of the QYBE. The quantum R-matrix is associated with a classical r-matrix r_{ij} through

$$R_{ij} = \exp\left(\hbar\, r_{ij}\right), \tag{2.94}$$

where $\hbar$ is a real constant.

Taking $\hbar$ to be infinitesimal, R_{ij} can be expanded as

$$R_{ij} \cong 1 + \hbar\, r_{ij} + O(\hbar^2). \tag{2.95}$$

Then the QYBE (2.93) can be expanded as follows:

$$\begin{aligned}
&(1 + \hbar\, r_{12})(1 + \hbar\, r_{13})(1 + \hbar\, r_{23}) + O(\hbar^3) \\
&= (1 + \hbar\, r_{23})(1 + \hbar\, r_{13})(1 + \hbar\, r_{12}) + O(\hbar^3).
\end{aligned} \tag{2.96}$$

The zero-th order and the first-order terms in $\hbar$ are trivially satisfied. At the second order of $\hbar$, a non-trivial relation is obtained as

$$[r_{12}, r_{13}] + [r_{13}, r_{23}] + [r_{12}, r_{23}] = 0. \tag{2.97}$$

This is nothing but the hCYBE.

Appendix 3: Derivation of (2.24)

In this Appendix, let us derive the equation of motion (2.24) from the deformed action (2.21).

By considering an infinitesimal variation $\delta g = g \cdot \epsilon$, the variation of the classical action is given by

$$\delta S^{(\eta)} = -\frac{1}{2}\int d^2x\, \mathrm{Tr}\left[(\partial_+ A_- + \partial_- A_+ - [A_+, A_-] - [A_-, J_+])\,\epsilon\right], \tag{2.98}$$

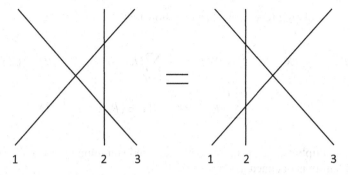

Fig. 2.1 QYBE means that these two diagrams are equivalent. Namely, the vertical line 2 can be moved freely horizontally. This situation can be realized in Feynmann diagrams in integrable quantum field theories due to the presence of infinite number of conserved charges

where we have dropped off the surface term.

By noting that

$$[A_+, J_-] = [A_+, A_-] + \eta[A_+, R(A_-)],$$
$$[A_-, J_+] = [A_-, A_+] - \eta[A_-, R(A_+)], \tag{2.99}$$

the variation of the action can be rewritten as

$$\delta S^{(\eta)} = -\frac{1}{2} \int d^2x \, \mathrm{Tr} \left[(\partial_+ A_- + \partial_- A_+ - \eta[A_+, R(A_-)] - \eta[R(A_+), A_-])\epsilon \right].$$

Thus the equation of motion (2.24) has been derived.

Appendix 4: Derivation of (2.64)

Here, let us derive the equation of motion (2.64) for the Yang–Baxter deformed symmetric coset sigma model (2.63).

To evaluate a variation of the projected current $\delta\tilde{A}$ with $\delta g = g\epsilon$, it is useful to prepare some identities. First of all, let us evaluate the variation of $R_g \circ P(X)$ as

$$\begin{aligned}
\delta \left(R_g \circ P(X) \right) &= \delta \left(g^{-1} R(g P(X) g^{-1}) g \right) \\
&= (-g^{-1}\delta g \cdot g^{-1}) R(g P(X) g^{-1}) g + g^{-1} R(g P(X) g^{-1}) \delta g \\
&\quad + g^{-1} R(g P(\delta X) g^{-1}) g \\
&\quad + g^{-1} R(\delta g P(X) g^{-1}) g + g^{-1} R(g P(X)(-g^{-1}\delta g \cdot g^{-1})) g \\
&= [R_g \circ P(X), \epsilon] + R_g \circ P(\delta X) - R_g([P(X), \epsilon]). \tag{2.100}
\end{aligned}$$

By using this relation repeatedly, one can show that

$$\delta\left(\left(R_g \circ P\right)^n (X)\right) = (R_g \circ P)^n (\delta X) + \sum_{k=0}^{n-1}(R_g \circ P)^k \left[(R_g \circ P)^{n-k}(X), \epsilon\right]$$

$$-\sum_{k=0}^{n-1}(R_g \circ P)^k R_g \left([P \circ (R_g \circ P)^{n-1-k}(X), \epsilon]\right). \quad (2.101)$$

Then, by multiplying $(\pm\eta)^n$ on the both sides and summing up n from 0 to ∞, the desired identity is obtained as

$$\delta\left(\frac{1}{1 \mp \eta R_g \circ P} X\right) \qquad\qquad\qquad\qquad\qquad (2.102)$$

$$= \frac{1}{1 \mp \eta R_g \circ P}\left(\delta X + \left[\frac{\pm\eta R_g \circ P}{1 \mp \eta R_g \circ P}X, \epsilon\right] \mp \eta R_g \left[P \frac{1}{1 \mp \eta R_g \circ P}X, \epsilon\right]\right).$$

Finally, substituting $J_\pm$ for X, the variation of $\tilde{A}_\pm$ is given by

$$\delta\tilde{A}_\pm = \frac{1}{1 \mp \eta R_g \circ P}\left(\partial_\pm\epsilon + [\tilde{A}_\pm, \epsilon] \mp \eta R_g[P(\tilde{A}_\pm), \epsilon]\right). \quad (2.103)$$

Now it is ready to evaluate the variation of the deformed action (2.63). Using the identity (2.103), one can show that

$$\delta S^{(\eta)} = -\frac{1}{2}\int d^2x \, \mathrm{Tr}\left[\left(\partial_+ P(\tilde{A}_-) + \partial_- P(\tilde{A}_+)\right)\right.$$

$$\left. +[\tilde{A}_+, P(\tilde{A}_-)] + [\tilde{A}_-, P(\tilde{A}_+)]\right)\epsilon\right], \quad (2.104)$$

and thus the classical equation of motion (2.64) has been derived.

Problems

2.1   A solution to hCYBE
Show that the R-operator (2.49) satisfies the hCYBE.

2.2   A variation of J_μ
Show the relation in (2.22).

2.3   A fundamental identity 1
Show the following relation:

$$\mathrm{Tr}\,(J_- A_+) = \mathrm{Tr}\,(A_- J_+),$$

where $A_\pm$ are defined in (2.20).

2.4   Rewrite $\mathcal{Z}$
Rewrite (2.25) into (2.27).

2.5   A fundamental identity 2
Show the following relation:

$$\text{Tr}\left(J_- P(\tilde{A}_+)\right) = \text{Tr}\left(P(\tilde{A}_-)J_+\right),$$

where $\tilde{A}_\pm$ are defined in (2.62).

References

1. I. Bena, J. Polchinski, R. Roiban, Hidden symmetries of the AdS$_5 \times$S^5 superstring. Phys. Rev. D **69**, 046002 (2004). [hep-th/0305116]
2. F. Delduc, M. Magro, B. Vicedo, On classical q-deformations of integrable sigma-models. JHEP **1311**, 192 (2013). arXiv:1308.3581 [hep-th]
3. F. Delduc, M. Magro, B. Vicedo, An integrable deformation of the AdS$_5 \times$S^5 superstring action. Phys. Rev. Lett. **112**(5), 051601 (2014). arXiv:1309.5850 [hep-th]
4. M. Hatsuda, K. Yoshida, Classical integrability and super Yangian of superstring on AdS$_5 \times$S^5. Adv. Theor. Math. Phys. **9**(5), 703 (2005) [hep-th/0407044]
5. M. Hatsuda, K. Yoshida, Super Yangian of superstring on AdS$_5 \times$S^5 revisited. Adv. Theor. Math. Phys. **15**(5), 1485 (2011). arXiv:1107.4673 [hep-th]
6. I. Kawaguchi, T. Matsumoto, K. Yoshida, Jordanian deformations of the AdS$_5 \times$S^5 superstring. JHEP **1404**, 153 (2014). arXiv:1401.4855 [hep-th]
7. C. Klimcik, Yang-Baxter sigma models and dS/AdS T duality. JHEP **0212**, 051 (2002). [hep-th/0210095]
8. C. Klimcik, On integrability of the Yang-Baxter sigma-model. J. Math. Phys. **50**, 043508 (2009). arXiv:0802.3518 [hep-th]
9. C. Klimcik, Integrability of the bi-Yang-Baxter sigma-model. Lett. Math. Phys. **104**, 1095 (2014). arXiv:1402.2105 [math-ph]
10. T. Matsumoto, K. Yoshida, Yang-Baxter sigma models based on the CYBE. Nucl. Phys. B **893**, 287 (2015). arXiv:1501.03665 [hep-th]

Chapter 3
Recent Progress on Yang–Baxter Deformation and Generalized Supergravity

Abstract In recent years, great progress has been made on a systematic method to perform an integrable deformation of a two-dimensional relativistic non-linear sigma model. The deformations are labeled by classical r-matrices satisfying the classical Yang–Baxter equation, and this method is called the Yang–Baxter deformation. It was generalized to type IIB superstring theory defined on the $AdS_5 \times S^5$ background and gave rise to a lot of integrable backgrounds including well-known backgrounds such as the Lunin-Maldacena background, a gravity dual for a non-commutative gauge theory, and a Schrödinger spacetime. In addition, the study of Yang–Baxter deformation led to the discovery of a generalized type IIB supergravity. In this chapter, I will give a short summary of the recent progress on the Yang–Baxter deformation and the generalized supergravity. For a comprehensive review, see [51].

3.1 Introduction

A conjectured duality between a string theory on the $(d + 1)$-dimensional anti de Sitter (AdS) space and a conformal field theory (CFT) in d dimensions, which is called the AdS/CFT correspondence (or simply AdS/CFT) [40], is one of the fascinating topics in String Theroy. A typical example of this correspondence is a duality between type IIB string theory defined on $AdS_5 \times S^5$ and the four-dimensional $\mathcal{N} = 4$ $SU(N)$ super Yang-Mills (SYM) theory in the large N limit.

One of the great achievements is the discovery of the integrable structure that exists behind AdS/CFT (For a comprehensive review, see [10]). As a tip of the iceberg of this integrable structure, type IIB superstring theory on $AdS_5 \times S^5$ [50], which is often abbreviated as the $AdS_5 \times S^5$ superstring, is classically integrable [11]. In the following, we will be concerned with this classical integrability.

An intriguing direction is to study an integrable deformation of the $AdS_5 \times S^5$ superstring. There are some possible ways in the context of integrable models, hence by employing one of them, one can perform an integrable deformation of the system as a two-dimensional non-linear sigma model. Accordingly, the target-space geometry is also deformed. The resulting background can be seen as a deformed $AdS_5 \times S^5$ geometry. Then, one may ask the following questions:

© The Author(s), under exclusive license to Springer Nature Singapore Pte Ltd. 2021   59
K. Yoshida, *Yang–Baxter Deformation of 2D Non-Linear Sigma Models*,
SpringerBriefs in Mathematical Physics,
https://doi.org/10.1007/978-981-16-1703-4_3

Does the deformation give a solution to type IIB supergravity or not?

If not, is the deformed background a solution to some new theory?

The main issue of this chapter is to answer these questions for a specific class of integrable deformations called the Yang–Baxter deformation [36, 37].

3.2   Yang–Baxter Deformation of the $AdS_5 \times S^5$ Superstring

Let us introduce Yang–Baxter deformation of the $AdS_5 \times S^5$ superstring [18, 19, 34]. To be pedagogical, we start from the explanation about the classical integrability of the $AdS_5 \times S^5$ superstring [11]. Then we introduce the Yang–Baxter deformed action and outline the supercoset construction. Finally, some examples are presented.

3.2.1   Classical Integrability of the $AdS_5 \times S^5$ Superstring

The classical action of the $AdS_5 \times S^5$ superstring is constructed based on the following supercoset [50]:

$$\frac{PSU(2, 2|4)}{SO(1, 4) \times SO(5)} . \tag{3.1}$$

The bosonic part of this coset describes the $AdS_5 \times S^5$ geometry,

$$\frac{SO(2, 4)}{SO(5)} \times \frac{SO(6)}{SO(5)} = AdS_5 \times S^5 . \tag{3.2}$$

The fermionic part of (3.1) corresponds to the spacetime fermions, whose dynamics is described in the Green-Schwarz (GS) formulation of superstring [50]. For a nice review on the $AdS_5 \times S^5$ superstring, see [7].

The bosonic part is nothing but a symmetric coset, hence the classical integrability of the system is ensured automatically. This symmetric coset structure is equivalent to the $\mathbb{Z}_2$-grading property. Remarkably, the supercoset (3.1) exhibits the $\mathbb{Z}_4$-grading as a supersymmetric generalization of the symmetric coset. This grading property ensures the classical integrability for the supersymmetric case, as elucidated by Bena, Polchinski and Roiban [11].

3.2.2   Yang–Baxter Deformed Action and Supercoset Construction

The Yang–Baxter deformation can also be applied to the $AdS_5 \times S^5$ superstring [18, 19, 34]. The deformed action is given by

$$S = -\frac{1}{2} \int_{-\infty}^{\infty} d\tau \int_0^{2\pi} d\sigma \; P_-^{ab} \text{Str} \left[J_a d \circ \frac{1}{1 - \eta R_g \circ d} (J_b) \right]. \tag{3.3}$$

When $\eta = 0$, the original Metsaev–Tseytlin action [50] is reproduced. For the detail of this action, see [18, 19, 34].

Since the deformed action (3.3) is written in terms of the group element, the target-space geometry is not clear. In addition, since the spacetime fermions are included, the dilaton and Ramond–Ramond (R-R) field strengths also appear as well as the metric and the Neveu-Schwarz–Neveu-Schwarz (NS-NS) two-form. In order to see the deformed background explicitly, one needs to perform supercoset construction by taking a parametrization of the group element [5, 6, 38]. Then, by expanding the action in terms of the spacetime fermion θ, the second-order action can be compared with the following canonical form of the GS superstring on an arbitrary background [15],

$$S = -\frac{\sqrt{\lambda_c}}{4} \int_{-\infty}^{\infty} d\tau \int_0^{2\pi} d\sigma \; [\gamma^{ab} G_{MN} \partial_a X^M \partial_b X^N - \epsilon^{ab} B_{MN} \partial_a X^M \partial_b X^N]$$
$$-\frac{\sqrt{\lambda_c}}{2} i \bar{\Theta}_I (\gamma^{ab} \delta^{IJ} - \epsilon^{ab} \sigma_3^{IJ}) e_a^m \Gamma_m D_b^{JK} \Theta_K + O(\theta^4). \tag{3.4}$$

This expression contains the metric G_{MN} and the NS–NS two-form B_{MN} manifestly. The covariant derivative D for the spacetime fermion θ is

$$D_a^{IJ} \equiv \delta^{IJ} \left(\partial_a - \frac{1}{4} \omega_a^{mn} \Gamma_{mn} \right) + \frac{1}{8} \sigma_3^{IJ} e_a^m H_{mnp} \Gamma^{np}$$
$$-\frac{1}{8} e^{\Phi} \left[\epsilon^{IJ} \Gamma^p F_p + \frac{1}{3!} \sigma_1^{IJ} \Gamma^{pqr} F_{pqr} + \frac{1}{2 \cdot 5!} \epsilon^{IJ} \Gamma^{pqrst} F_{pqrst} \right] e_a^m \Gamma_m$$

and it contains the dilaton Φ, and the R–R-field strengths F_p, F_{pqr} and F_{pqrst}. Thus, one can read off all of the (bosonic) components of type IIB supergravity from (3.4).

Here we should go back to the original questions made in Introduction. In principle, a new deformed background can be obtained by performing the supercoset construction with a classical r-matrix. Then the question can be rephrased as follows:

Are the resulting backgrounds solutions of type IIB supergravity?

The answer depends on classical r-matrices utilized as the initial input. Now we know the significant condition for this issue, which is called the unimodularity condition [12].

The unimodularity condition [12] is given by

$$r^{ij}[b_i, b_j] = 0 \qquad (r = r^{ij} b_i \wedge b_j \in \mathfrak{g} \otimes \mathfrak{g}). \tag{3.5}$$

When a classical r-matrix satisfies this condition, the resulting background is a solution to type IIB supergravity. If not, the background does not satisfy the on-shell condition of the supergravity and becomes a solution to a *generalized* supergravity. In the next section, we will explain what the generalized supergravity is. Before concluding this section, we will present some unimodular examples, which are well-known examples in different contexts (For short reviews, see [45, 49]).

3.2.3 Unimodular Examples

(1) Gamma-Deformation of S^5
A simple unimodular r-matrix is given by [43]

$$r = \frac{1}{8} \left(\mu_3\, h_1 \wedge h_2 + \mu_1\, h_2 \wedge h_3 + \mu_2\, h_3 \wedge h_1 \right), \tag{3.6}$$

where h_i $(i = 1, 2, 3)$ are the Cartan generators of $\mathfrak{su}(4)$ and μ_i $(i = 1, 2, 3)$ are constant parameters.

Then, the supercoset construction [38] leads to the following background:

$$
\begin{aligned}
ds^2 &= ds_{\mathrm{AdS}_5}^2 + \sum_{i=1}^{3}(d\rho_i^2 + G\rho_i^2 d\phi_i^2) + \eta^2 G \rho_1^2 \rho_2^2 \rho_3^2 \left(\sum_{i=1}^{3} \mu_i\, d\phi_i \right)^2, \\
B_2 &= \eta\, G \left(\mu_3\, \rho_1^2 \rho_2^2\, d\phi_1 \wedge d\phi_2 + \mu_1\, \rho_2^2 \rho_3^2\, d\phi_2 \wedge d\phi_3 + \mu_2\, \rho_3^2 \rho_1^2\, d\phi_3 \wedge d\phi_1 \right), \\
F_5 &= 4 \left[\omega_{\mathrm{AdS}_5} + G\, \omega_{S^5} \right], \qquad \Phi = \frac{1}{2} \log G, \\
F_3 &= -4\eta \sin^3 \alpha\, \cos\alpha\, \sin\theta\, \cos\theta \left(\sum_{i=1}^{3} \mu_i\, d\phi_i \right) \wedge d\alpha \wedge d\theta.
\end{aligned}
\tag{3.7}
$$

Here the scalar function G is given by

$$G^{-1} \equiv 1 + \eta^2 (1 + \mu_3^2 \rho_1^2 \rho_2^2 + \mu_1^2 \rho_2^2 \rho_3^2 + \mu_2^2 \rho_3^2 \rho_1^2), \qquad \sum_{i=1}^{3} \rho_i^2 = 1, \tag{3.8}$$

where ρ_i's are parametrized as

$$\rho_1 = \sin \alpha \cos \theta, \quad \rho_2 = \sin \alpha \sin \theta, \quad \rho_3 = \cos \alpha. \tag{3.9}$$

This is the gamma-deformation of S^5 presented in [23, 39]. Indeed, the classical r-matrix corresponds to applying so-called TsT transformations three times.

(2) Gravity Dual for Non-commutative Gauge Theory

Next, let us consider the following classical r-matrix [44],

$$r = \frac{1}{2} p_2 \wedge p_3 . \tag{3.10}$$

Here the generators are represented by

$$p_\mu \equiv \frac{1}{2} \gamma_\mu - m_{\mu 5} , \qquad m_{\mu 5} \equiv \frac{1}{4} [\gamma_\mu, \gamma_5] , \tag{3.11}$$

where γ_μ's are matrices of $\mathfrak{su}(2, 2)$.

Then the supercoset construction [38] gives rise to the background:

$$ds^2 = \frac{1}{z^2}(-dx_0^2 + dx_1^2) + \frac{z^2}{z^4 + \eta^2}(dx_2^2 + dx_3^2) + \frac{dz^2}{z^2} + d\Omega_5^2 ,$$

$$B_2 = \frac{\eta}{z^4 + \eta^2} dx^2 \wedge dx^3 , \qquad \Phi = \frac{1}{2} \log \left(\frac{z^4}{z^4 + \eta^2} \right) ,$$

$$F_3 = \frac{4\eta}{z^5} dx^0 \wedge dx^1 \wedge dz , \qquad F_5 = 4 \left[e^{2\Phi} \omega_{AdS_5} + \omega_{S^5} \right] . \tag{3.12}$$

This is nothing but a gravity dual of a noncommutative gauge theory[1] constructed in [25, 41]. The classical r-matrix (3.10) also corresponds to a TsT tranformation.

(3) Schrödinger Spacetime

The last one is composed of the generators of both $\mathfrak{su}(2, 2)$ and $\mathfrak{su}(4)$ [48]:

$$r = -\frac{i}{4} p_- \wedge (h_1 + h_2 + h_3) , \tag{3.13}$$

where the above generators have already appeared.

By the supercoset construction [38], the resulting background is given by

$$ds^2 = \frac{-2dx^+ dx^- + (dx^1)^2 + (dx^2)^2 + dz^2}{z^2} - \eta^2 \frac{(dx^+)^2}{z^4} + ds_{S^5}^2$$

$$B_2 = \frac{\eta}{z^2} dx^+ \wedge (d\chi + \omega) , \qquad \Phi = \text{const.} ,$$

$$F_5 = 4 \left[e^{2\Phi} \omega_{AdS_5} + \omega_{S^5} \right] , \tag{3.14}$$

where the S^5-coordinates are taken as

$$ds_{S^5}^2 = (d\chi + \omega)^2 + ds_{\mathbb{CP}^2}^2 , \tag{3.15}$$

$$ds_{\mathbb{CP}^2}^2 = d\mu^2 + \sin^2 \mu \left(\Sigma_1^2 + \Sigma_2^2 + \cos^2 \mu \, \Sigma_3^2 \right) . \tag{3.16}$$

---

[1]This means a gauge theory defined on a noncommutative spacetime [60].

This is the 5D Schrödinger spacetime embedded in type IIB supergravity [1, 28, 42]. The classical r-matrix (3.13) corresponds to a null Melvin twist.

In fact, the three examples presented so far belong to the class of abelian classical r-matrix. All of the Yang–Baxter deformations in this class can be expressed as TsT transformations [56]. For more general cases, see [12, 58].

3.3 Generalized Supergravity

In this section, let us introduce an extension of type IIB supergravity, called the generalized supergravity. The bosonic part was discovered originally by Arutunov et al. [8] in the study of Yang–Baxter deformation of the $AdS_5 \times S^5$ superstring. After that, Tseytlin and Wulff succeeded in reproducing the generalized supergravity including the fermionic sector (dilatino and gravitino) by solving the kappa-symmetry constraints of the GS formulation of type IIB superstring on an arbitrary background [61].

The equations of motion in (the bosonic sector of) the generalized supergravity are given by

$$R_{MN} - \frac{1}{4} H_{MKL} H_N{}^{KL} - T_{MN} + D_M X_N + D_N X_M = 0, \tag{3.17}$$

$$\frac{1}{2} D^K H_{KMN} + \frac{1}{2} F^K F_{KMN} + \frac{1}{12} F_{MNKLP} F^{KLP} \tag{3.18}$$

$$= X^K H_{KMN} + D_M X_N - D_N X_M, \tag{3.19}$$

$$R - \frac{1}{12} H^2 + 4 D_M X^M - 4 X_M X^M = 0, \tag{3.20}$$

$$D^M \mathcal{F}_M - Z^M \mathcal{F}_M - \frac{1}{6} H^{MNK} \mathcal{F}_{MNK} = 0, \qquad I^M \mathcal{F}_M = 0, \tag{3.21}$$

$$D^K \mathcal{F}_{KMN} - Z^K \mathcal{F}_{KMN} - \frac{1}{6} H^{KPQ} \mathcal{F}_{KPQMN} - (I \wedge \mathcal{F}_1)_{MN} = 0, \tag{3.22}$$

$$D^K \mathcal{F}_{KMNPQ} - Z^K \mathcal{F}_{KMNPQ}$$

$$+ \frac{1}{36} \epsilon_{MNPQRSTUVW} H^{RST} \mathcal{F}^{UVW} - (I \wedge \mathcal{F}_3)_{MNPQ} = 0. \tag{3.23}$$

Here $H_3 \equiv d B_2$ is the field-strength for NS-NS two-form B_2 and $\mathcal{F}_{p+1} \equiv e^{\Phi} F_{p+1}$ where $F_{p+1} \equiv d C_p$ is the field-strength for R-R p-form C_p, and Φ is the dilaton. The energy-momentum tensor T_{MN} in (3.17) is given by

$$T_{MN} \equiv \frac{1}{2} \mathcal{F}_M \mathcal{F}_N + \frac{1}{4} \mathcal{F}_{MKL} \mathcal{F}_N{}^{KL} + \frac{1}{4 \times 4!} \mathcal{F}_{MPQRS} \mathcal{F}_N{}^{PQRS}$$

$$- \frac{1}{4} G_{MN} (\mathcal{F}_K \mathcal{F}^K + \frac{1}{6} \mathcal{F}_{PQR} \mathcal{F}^{PQR}). \tag{3.24}$$

The modified parts are three vector fields X_M, I_M and Z_M. But $X_M = I_M + Z_M$. So two of them, say I_M and Z_M are independent fields. Note that Z_M was originally the derivative of dilaton but now has undergone some modification.

Now the Bianchi identities are also modified as

$$(d\mathcal{F}_1 - Z \wedge \mathcal{F}_1)_{MN} - I^K \mathcal{F}_{MNK} = 0,$$
$$(d\mathcal{F}_3 - Z \wedge \mathcal{F}_3 + H_3 \wedge \mathcal{F}_1)_{MNPQ} - I^K \mathcal{F}_{MNPQK} = 0, \qquad (3.25)$$
$$(d\mathcal{F}_5 - Z \wedge \mathcal{F}_5 + H_3 \wedge \mathcal{F}_3)_{MNPQRS} + \frac{1}{6}\epsilon_{MNPQRSTUVW} I^T \mathcal{F}^{UVW} = 0.$$

Furthermore, we need to explain more constraints,

$$D_M I_N + D_N I_M = 0, \qquad (3.26)$$
$$D_M Z_N - D_N Z_M + I^K H_{KMN} = 0, \qquad (3.27)$$
$$I^M Z_M = 0. \qquad (3.28)$$

In particular, the first condition (3.26) is nothing but the Killing condition for I. Namely, I should be taken as a Killing vector. This condition may sound a bit stronger but this condition is necessary in solving the kappa-symmetry constraints [61]. Furthermore, this Killing condition is necessary to consider the embedding of the generalized supergravity into Double Field Theory [9, 57, 59].

The Lie derivative of NS-NS two-form B_2 along the Killing direction

$$(\mathcal{L}_I B)_{MN} = I^K \partial_K B_{MN} + B_{KN} \partial_M I^K - B_{KM} \partial_N I^K$$

should vanish. Then, by solving the second condition (3.27), one can obtain the following expression:

$$Z_M = \partial_M \Phi - B_{MN} I^N.$$

From this expression, one can understand that Z_M is a modification of the dilaton derivative with non-vanishing I and that Z is not independent of I. In this sense, only the Killing vector I characterizes the generalized supergravity. When $I = 0$, the original type IIB supergravity is reproduced.

3.3.1 Non-unimodular Example

As denoted previously, a non-unimodular classical r-matrix leads to a solution to the generalized supergravity with $I \neq 0$.

Let consider here the following non-unimodular example [35, 46]:

$$r = E_{24} \wedge (c_1 E_{22} - c_2 E_{44})$$
$$= (p_0 - p_3) \wedge \left[a_1 \left(\frac{1}{2} \gamma_5 - n_{03} \right) - a_2 \left(n_{12} - \frac{i}{2} \mathbf{1}_4 \right) \right], \tag{3.29}$$

where this is a two-parameter family and the deformation parameters (c_1, c_2) are related to (a_1, a_2) through the relation

$$a_1 \equiv \frac{c_1 + c_2}{2} = \mathrm{Re}(c_1), \qquad a_2 \equiv \frac{c_1 - c_2}{2i} = \mathrm{Im}(c_1). \tag{3.30}$$

Then, by performing the supercoset construction [38], one can obtain the following background:

$$ds^2 = \frac{-2dx^+ dx^- + d\rho^2 + \rho^2 d\phi^2 + dz^2}{z^2} + ds_{S^5}^2$$
$$-4\eta^2 \left[(a_1^2 + a_2^2) \frac{\rho^2}{z^6} + \frac{a_1^2}{z^4} \right] (dx^+)^2,$$

$$B_2 = 8\eta \left[\frac{a_1 x^1 + a_2 x^2}{z^4} dx^+ \wedge dx^1 + \frac{a_1 x^2 - a_2 x^1}{z^4} dx^+ \wedge dx^2 \right.$$
$$\left. + a_1 \frac{1}{z^3} dx^+ \wedge dz \right],$$

$$F_3 = 8\eta \left[\frac{a_2 x^1 - a_1 x^2}{z^5} dx^+ \wedge dx^1 \wedge dz + \frac{a_1 x^1 + a_2 x^2}{z^5} dx^+ \wedge dx^2 \wedge dz \right.$$
$$\left. + \frac{a_1}{z^4} dx^+ \wedge dx^1 \wedge dx^2 \right],$$

$$F_5 = \text{undeformed}, \qquad \Phi = \text{const.} \tag{3.31}$$

This background is not a solution to type IIB supergravity. It is easy to check this statement by taking an exterior derivative of F_3,

$$dF_3 = 16\eta \frac{a_1}{z^5} dx^+ \wedge dx^1 \wedge dx^2 \wedge dz \neq 0. \tag{3.32}$$

This does not vanish and the equation of motion for B_2 is also not satisfied.

However, by taking the extra vector field I as

$$I = -\frac{2\eta a_1}{z^2} dx^+, \qquad Z = 0,$$

the background (3.31) becomes a solution to the generalized supergravity [38]. For other non-unimodular solutions, for example, see [29, 52].

3.3.2 Hoare-Tseytlin Conjecture

What of the generalized supergravity is so interesting? In the long history that String Theory has been studied, a number of so-called "pathological backgrounds," which are not solutions to supergravities, have been discovered. For example, it is well-known that non-abelian T-dualities generate such pathological backgrounds. It may be a good idea to check whether these backgrounds may be solutions to the generalized supergravity.

In fact, Hoare and Tseytlin advocated an interesting conjecture, the homogeneous Yang–Baxter deformations are equivalent to (a certain class of) non-abelian T-dualities [30]. Then this conjecture was proven by Borsato and Wulff [13]. The Yang–Baxter deformed backgrounds are solutions to the generalized supergravity, hence the accompanying non-abelian T-dualized backgrounds are also solutions as well.

3.3.3 Non Yang–Baxter Solution

As we have seen so far, the Yang–Baxter deformation can be regarded as a solution generation technique in the generalized supergravity. However, as a matter of course, it does not give all of the solutions. That is, there exist a number of solutions which cannot be obtained as Yang–Baxter deformations.

Such an example is the Gasperini–Ricci–Veneziano background [24]:

$$ds^2 = -dt^2 + \frac{(t^4 + y^2)\, dx^2 - 2x\, y\, dx\, dy + (t^4 + x^2)\, dy^2 + t^4 dz^2}{t^2(t^4 + x^2 + y^2)} + ds_{T^6}^2 \,,$$

$$B_2 = \frac{(x\, dx + y\, dy) \wedge dz}{t^4 + x^2 + y^2} \,, \qquad \Phi = \frac{1}{2} \ln\left[\frac{1}{t^2(t^4 + x^2 + y^2)}\right] . \tag{3.33}$$

This is not a solution of the usual supergravity. However, by taking the vector field I like

$$I^z = -2 \,,$$

the background (3.33) becomes a solution to the generalized supergravity [21]. Note here that this background can be obtained through a non-abelian T-duality, but cannot be expressed as a Yang–Baxter deformation. Hence this background is not included in the Hoare-Tseytlin conjecture. The background (3.33) is just an example, but further confirmation was made in [32], in which a number of similar solutions were listed.

3.4   Other Topics

Due to the page limit, a number of other issues could not be covered here. The list of them includes

- Open string picture, non-commutativity and Killing spinor formula [2–4, 31, 44, 53, 54, 62–64]
- Embedding of the generalized supergravity into Double Field Theory (DFT) and the DFT perspective [9, 26, 27, 55, 57, 59].
- Non-geometric backgrounds obtained as Yang–Baxter deformations [21]
- Arguments on Weyl invariance of string theory on a generalized supergravity background [22, 57]
- Relation between Costello–Yamazaki [14] and Yang–Baxter deformation [16]

and more. I apologize for not being able to make a complete list, and hope that I have the opportunity to write a more comprehensive book.

References

1. A. Adams, K. Balasubramanian, J. McGreevy, Hot spacetimes for cold atoms. JHEP **11**, 059 (2008). arXiv:0807.1111 [hep-th]
2. T. Araujo, I. Bakhmatov, E. Ó Colgáin, J. Sakamoto, M.M. Sheikh-Jabbari, K. Yoshida, Yang–Baxter σ-models, conformal twists, and noncommutative Yang–Mills theory. Phys. Rev. **D95**(10), 105006 (2017). arXiv:1702.02861 [hep-th]
3. T. Araujo, I. Bakhmatov, E. Ó Colgáin, J. Sakamoto, M.M. Sheikh-Jabbari, K. Yoshida, Conformal twists, Yang–Baxter σ-models & holographic noncommutativity. J. Phys. A **51**(23), 235401 (2018). arXiv:1705.02063 [hep-th]
4. T. Araujo, E. Ó Colgáin, J. Sakamoto, M.M. Sheikh-Jabbari, K. Yoshida, I in generalized supergravity. Eur. Phys. J. C **77**(11), 739 (2017). arXiv:1708.03163 [hep-th]
5. G. Arutyunov, R. Borsato, S. Frolov, S-matrix for strings on η-deformed AdS$_5 \times$S^5. JHEP **04**, 002 (2014). arXiv:1312.3542 [hep-th]
6. G. Arutyunov, R. Borsato, S. Frolov, Puzzles of η-deformed AdS$_5 \times$S^5. JHEP **12**, 049 (2015). arXiv:1507.04239 [hep-th]
7. G. Arutyunov, S. Frolov, Foundations of the AdS$_5 \times$S^5 superstring, I. J. Phys. A **42**, 254003 (2009). arXiv:0901.4937 [hep-th]
8. G. Arutyunov, S. Frolov, B. Hoare, R. Roiban, A.A. Tseytlin, Scale invariance of the η-deformed $AdS_5 \times S^5$ superstring, T-duality and modified type II equations. Nucl. Phys. B **903**, 262–303 (2016). arXiv:1511.05795 [hep-th]
9. A. Baguet, M. Magro, H. Samtleben, Generalized IIB supergravity from exceptional field theory. JHEP **03**, 100 (2017). arXiv:1612.07210 [hep-th]
10. N. Beisert et al., Review of AdS/CFT integrability: an overview. Lett. Math. Phys. **99**, 3–32 (2012). arXiv:1012.3982 [hep-th]
11. I. Bena, J. Polchinski, R. Roiban, Hidden symmetries of the AdS$_5 \times$S^5 superstring. Phys. Rev. D **69**, 046002 (2004). [hep-th/0305116]
12. R. Borsato, L. Wulff, Target space supergeometry of η and λ-deformed strings. JHEP **10**, 045 (2016). arXiv:1608.03570 [hep-th]
13. R. Borsato, L. Wulff, Integrable deformations of T-dual σ models. Phys. Rev. Lett. **117**(25), 251602. arXiv:1609.09834 [hep-th]
14. K. Costello, M. Yamazaki, Gauge theory and integrability, III. arXiv:1908.02289 [hep-th]

15. M. Cvetic, H. Lu, C.N. Pope, K.S. Stelle, T duality in the Green–Schwarz formalism, and the massless/massive IIA duality map. Nucl. Phys. B **573**, 149 (2000). [hep-th/9907202]
16. F. Delduc, S. Lacroix, M. Magro, B. Vicedo, A unifying 2d action for integrable σ-models from 4d Chern-Simons theory. arXiv:1909.13824 [hep-th]
17. F. Delduc, M. Magro, B. Vicedo, On classical q-deformations of integrable sigma-models. JHEP **11**, 192 (2013). arXiv:1308.3581 [hep-th]
18. F. Delduc, M. Magro and B. Vicedo, An integrable deformation of the $AdS_5 \times S^5$ superstring action. Phys. Rev. Lett. **112**(5), 051601 (2014). arXiv:1309.5850 [hep-th]
19. F. Delduc, M. Magro, B. Vicedo, Derivation of the action and symmetries of the q-deformed $AdS_5 \times S^5$ superstring. JHEP **10**, 132 (2014). arXiv:1406.6286 [hep-th]
20. V. G. Drinfeld, Hopf algebras and the quantum Yang-Baxter equation. Sov. Math. Dokl. **32**, 254–258 (1985). [Dokl. Akad. Nauk Ser. Fiz. **283**, 1060 (1985)]
21. J.J. Fernández-Melgarejo, J. Sakamoto, Y. Sakatani, K. Yoshida, T-folds from Yang–Baxter deformations. JHEP **1712**, 108 (2017). arXiv:1710.06849 [hep-th]
22. J.J. Fernández-Melgarejo, J. Sakamoto, Y. Sakatani, K. Yoshida, Weyl invariance of string theories in generalized supergravity backgrounds. Phys. Rev. Lett. **122**(11), 111602 (2019). arXiv:1811.10600 [hep-th]
23. S. Frolov, Lax pair for strings in Lunin–Maldacena background. JHEP **05**, 069 (2005). [hep-th/0503201]
24. M. Gasperini, R. Ricci, G. Veneziano, A problem with nonAbelian duality? Phys. Lett. B **319**, 438–444 (1993). [hep-th/9308112]
25. A. Hashimoto, N. Itzhaki, Noncommutative Yang–Mills and the AdS/CFT correspondence. Phys. Lett. B **465**, 142–147 (1999). [hep-th/9907166]
26. F. Hassler, The topology of double field theory. JHEP **1804**, 128 (2018). arXiv:1611.07978 [hep-th]
27. F. Hassler, Poisson-Lie T-Duality in double field theory. arXiv:1707.08624 [hep-th]
28. C.P. Herzog, M. Rangamani, S.F. Ross, Heating up Galilean holography. JHEP **11**, 080 (2008). arXiv:0807.1099 [hep-th]
29. B. Hoare, S.J. van Tongeren, On jordanian deformations of AdS_5 and supergravity. J. Phys. **A49**(43), 434006 (2016). arXiv:1605.03554 [hep-th]
30. B. Hoare, A. A. Tseytlin, Homogeneous Yang–Baxter deformations as non-abelian duals of the AdS_5 sigma-model. J. Phys. **A49**(49), 494001 (2016). arXiv:1609.02550 [hep-th]
31. B. Hoare, D.C. Thompson, Marginal and non-commutative deformations via non-abelian T-duality. JHEP **02**, 059 (2017). arXiv:1611.08020 [hep-th]
32. M. Hong, Y. Kim and E. Ó Colgáin, On non-Abelian T-duality for non-semisimple groups. Eur. Phys. J. C **78**(12), 1025 (2018). arXiv:1801.09567 [hep-th]
33. M. Jimbo, A q difference analog of $U(g)$ and the Yang–Baxter equation. Lett. Math. Phys. **10**, 63–69 (1985)
34. I. Kawaguchi, T. Matsumoto, K. Yoshida, Jordanian deformations of the $AdS_5 \times S^5$ superstring. JHEP **04**, 153 (2014). arXiv:1401.4855 [hep-th]
35. I. Kawaguchi, T. Matsumoto, K. Yoshida, A Jordanian deformation of AdS space in type IIB supergravity. JHEP **06**, 146 (2014). arXiv:1402.6147 [hep-th]
36. C. Klimcik, Yang–Baxter sigma models and dS/AdS T duality. JHEP **12**, 051 (2002). [hep-th/0210095]
37. C. Klimcik, On integrability of the Yang–Baxter sigma-model. J. Math. Phys. **50**, 043508 (2009). arXiv:0802.3518 [hep-th]
38. H. Kyono, K. Yoshida, Supercoset construction of Yang–Baxter deformed $AdS_5 \times S^5$ backgrounds. PTEP **2016**(8), 083B03 (2016). arXiv:1605.02519 [hep-th]
39. O. Lunin, J.M. Maldacena, Deforming field theories with $U(1) \times U(1)$ global symmetry and their gravity duals. JHEP **05**, 033 (2005). [hep-th/0502086]
40. J.M. Maldacena, The large N limit of superconformal field theories and supergravity. Int. J. Theor. Phys. **38**, 1113–1133 (1999). arXiv:hep-th/9711200 [hep-th] [Adv. Theor. Math. Phys. 2, 231 (1998)]

41. J.M. Maldacena, J.G. Russo, Large N limit of noncommutative gauge theories. JHEP **09**, 025 (1999). [hep-th/9908134]

42. J. Maldacena, D. Martelli, Y. Tachikawa, Comments on string theory backgrounds with non-relativistic conformal symmetry. JHEP **10**, 072 (2008). arXiv:0807.1100 [hep-th]

43. T. Matsumoto, K. Yoshida, Lunin-Maldacena backgrounds from the classical Yang–Baxter equation—towards the gravity/CYBE correspondence. JHEP **06**, 135 (2014). arXiv:1404.1838 [hep-th]

44. T. Matsumoto, K. Yoshida, Integrability of classical strings dual for noncommutative gauge theories. JHEP **06**, 163 (2014). arXiv:1404.3657 [hep-th]

45. T. Matsumoto, K. Yoshida, Integrable deformations of the $AdS_5 \times S^5$ superstring and the classical Yang–Baxter equation-Towards the gravity/CYBE correspondence. J. Phys. Conf. Ser. **563**(1), 012020 (2014). arXiv:1410.0575 [hep-th]

46. T. Matsumoto, K. Yoshida, Yang–Baxter deformations and string dualities. JHEP **03**, 137 (2015). arXiv:1412.3658 [hep-th]

47. T. Matsumoto, K. Yoshida, Yang–Baxter sigma models based on the CYBE. Nucl. Phys. B **893**, 287–304 (2015). arXiv:1501.03665 [hep-th]

48. T. Matsumoto, K. Yoshida, Schrödinger geometries arising from Yang–Baxter deformations. JHEP **04**, 180 (2015). arXiv:1502.00740 [hep-th]

49. T. Matsumoto, K. Yoshida, Towards the gravity/CYBE correspondence-the current status. J. Phys. Conf. Ser. **670**(1), 012033 (2016)

50. R.R. Metsaev, A.A. Tseytlin, Type IIB superstring action in $AdS_5 \times S^5$ background. Nucl. Phys. B **533**, 109–126 (1998). [hep-th/9805028]

51. D. Orlando, S. Reffert, J. Sakamoto, Y. Sekiguchi, K. Yoshida, Yang-Baxter deformations and generalized supergravity—A short summary. arXiv:1912.02553 [hep-th]

52. D. Orlando, S. Reffert, J. Sakamoto, K. Yoshida, Generalized type IIB supergravity equations and non-Abelian classical r-matrices. J. Phys. **A49**(44), 445403 (2016). arXiv:1607.00795 [hep-th]

53. D. Orlando, S. Reffert, Y. Sekiguchi, K. Yoshida, Killing spinors from classical r-matrices. J. Phys. A **51**(39), 395401 (2018). arXiv:1805.00948 [hep-th]

54. D. Orlando, S. Reffert, Y. Sekiguchi and K. Yoshida, "SUSY and the bi-vector," Phys. Scripta **94** (2019) no. 9, 095001 arXiv:1811.11764 [hep-th]

55. D. Orlando, S. Reffert, Y. Sekiguchi, K. Yoshida, $O(d, d)$ transformations preserve classical integrability. Nucl. Phys. B **950**, 114880 (2020). arXiv:1907.03759 [hep-th]

56. D. Osten, S.J. van Tongeren, Abelian Yang–Baxter deformations and TsT transformations. Nucl. Phys. B **915**, 184–205 (2017). arXiv:1608.08504 [hep-th]

57. J. Sakamoto, Y. Sakatani, K. Yoshida, Weyl invariance for generalized supergravity backgrounds from the doubled formalism. PTEP **2017**(5), 053B07 (2017). arXiv:1703.09213 [hep-th]

58. J. Sakamoto, Y. Sakatani, K. Yoshida, Homogeneous Yang–Baxter deformations as generalized diffeomorphisms. J. Phys. **A50**(41), 415401 (2017). arXiv:1705.07116 [hep-th]

59. Y. Sakatani, S. Uehara, K. Yoshida, Generalized gravity from modified DFT. JHEP **04**, 123 (2017). arXiv:1611.05856 [hep-th]

60. N. Seiberg, E. Witten, String theory and noncommutative geometry. JHEP **09**, 032 (1999). [hep-th/9908142]

61. A.A. Tseytlin, L. Wulff, Kappa-symmetry of superstring sigma model and generalized 10d supergravity equations. JHEP **06**, 174 (2016). arXiv:1605.04884 [hep-th]

62. S.J. van Tongeren, On classical Yang–Baxter based deformations of the $AdS_5 \times S^5$ superstring. JHEP **06**, 048 (2015). arXiv:1504.05516 [hep-th]

63. S.J. van Tongeren, Yang–Baxter deformations, AdS/CFT, and twist-noncommutative gauge theory. Nucl. Phys. B **904**, 148–175 (2016). arXiv:1506.01023 [hep-th]

64. S.J. van Tongeren, Almost abelian twists and AdS/CFT. Phys. Lett. B **765**, 344–351 (2017). arXiv:1610.05677 [hep-th]

Printed in the United States
by Baker & Taylor Publisher Services